CHEMICAL REACTION MECHANISMS

Mechanisms of Oxidation Reactions

Volume 1

Oxygen Atom Transfer Reactions

Edited by

Robert Bakhtchadjian

Institute of Chemical Physics
National Academy of Sciences of the Republic of Armenia
Yerevan
Republic of Armenia

CHEMICAL RECTION MECHANISMS

Mechanisms of Oxidation Reactions

Volume 1

Oxygen Atom Transfer Reactions

Editor: Robert Bakhtchadjian

ISBN (Online): 978-981-5050-92-9

ISBN (Print): 978-981-5050-93-6

ISBN (Paperback): 978-981-5050-94-3

© 2023, Bentham Books imprint.

Published by Bentham Science Publishers Pte. Ltd. Singapore. All Rights Reserved.

First published in 2023.

need for a court order if at any point you breach any terms of this License Agreement. In no event will any delay or failure by Bentham Science Publishers in enforcing your compliance with this License Agreement constitute a waiver of any of its rights.

3. You acknowledge that you have read this License Agreement, and agree to be bound by its terms and conditions. To the extent that any other terms and conditions presented on any website of Bentham Science Publishers conflict with, or are inconsistent with, the terms and conditions set out in this License Agreement, you acknowledge that the terms and conditions set out in this License Agreement shall prevail.

Bentham Science Publishers Pte. Ltd.
80 Robinson Road #02-00
Singapore 068898
Singapore
Email: subscriptions@benthamscience.net

CONTENTS

FOREWORD

Oxygen atom transfer reactions have been widely explored in biological systems and chemical synthesis studies. This volume outlines some basic mechanistic understanding and recent achievements in the study of the oxygen atom transfer reactions catalyzed by transition metal-oxo complexes. The influence of the nature of the central metals and the coordinated ligands on mechanisms of oxygen atom transfer reactions are summarized. In the first chapter, the type of oxidation reactions and the general classification of oxygen atom transfer reactions are introduced in detail. It is very useful for readers to be acquainted with the catalytic oxidation reaction via oxygen atom transfer of organometallic complexes. The following chapters of this volume are reviews involving the study of different aspects of oxygen atom transfer reaction mechanisms, such as O_2 activation driven by transition metal complexes and oxygen atom transfer reactions catalyzed by nickel-based organic complexes. These chapters provide readers with some efficient catalytic strategies for the activation of O_2 and the functionalization of C-H bonds and C=C bonds. This volume not only offers the basic knowledge of oxygen atom transfer reactions but also introduces the main development of this field. This book is promising to play an important role in motivating the interests of chemists and biology scientists from all over the world to further develop oxygen atom transfer reactions. The knowledge obtained in this field should also serve other oxidation reactions.

Jincai Zhao
Professor, Ph.D.
Member of the European Academy of Sciences
Member of the Chinese Academy of Sciences
People's Republic of China

SERIES PREFACE

The investigation of the reaction mechanism plays a central role in chemistry. The overall chemical transformation of substances is a complex process that often involves elementary chemical reactions, the sequence of which composes the reaction mechanism. Modern perceptions of the mechanisms of chemical reactions are based on both experimental and theoretical investigations in physics, chemistry, and biology. The study of reactions and their mechanisms requires periodic adjustment, detailing, and permanent perfection in the light of new experimental data and theoretical perceptions. The discovery of new reactions and investigation of their kinetic peculiarities change the perceptions of the existing reaction mechanisms. Sometimes, over a long period, perceptions of how the reaction occurs can be changed so much that only the historical significance of their initial version of the mechanism may remain in science. The introduction of new ideas and new concepts in science and the changes related to the reaction mechanism is a permanent process. However, usually, this information is scattered across various specialized periodicals and scientific reports. It is clear that for a certain period, it becomes necessary to collect and summarize information about the reaction mechanisms in more general editions in the form of book series.

In this context, the aim of the creation of this Book Series is to present a certain part of the modern achievements in mechanism investigations in some important fields of chemistry and biology. The mechanisms of various classes of chemical reactions will be the subject of separate volumes of this book series. The first two volumes are devoted to the mechanisms of oxidation reactions.

The first volume is entitled as:

Mechanisms of Oxidation Reactions: Volume 1. Oxygen Atom Transfer Reactions

I am grateful to Bentham Science for this opportunity to create and edit this Book Series.

Robert Bakhtchadjian
Institute of Chemical Physics
National Academy of Sciences of the Republic of Armenia
Yerevan, Republic of Armenia

PREFACE

Understanding the reaction mechanism is one of the keys to achieving controllable processes in chemistry, biology, and some applied sciences. Obviously, this refers to oxidation processes that are so widespread in nature, including living cells, and in manmade systems, including the chemical industry. Oxidation processes in the chemical industry are mainly catalytic reactions, in general, using metals, their oxides or organometallic compounds as catalysts. In living systems, enzymes, natural complex catalysts, also containing metallic elements, most of which are organometallic complexes of transition metals play a similar role. From the viewpoint of the reaction mechanism, the reactions occurring *via* the transfer of oxygen atoms to the substrate are one of the widespread types of oxidation processes, observed both in manmade chemical and natural biological systems. Currently, the high efficiency and selectivity of enzymatic oxidation under very mild conditions are not yet available in manmade chemical systems. What and how can we learn from Nature? Two very close and, at the same time, different approaches may serve this purpose. They are known as bio-inspiration and biomimicry. In this regard, the present volume, discussing different catalytic strategies, also involves certain achievements obtained both in bio-inspired and biomimetic systems in comparison with the application of traditional organometallic catalysts of transition metal elements in oxidation reactions.

The intended audience of this book may comprise not only researchers in the fields of chemistry, physics, and biology, but also practitioners in the fields of chemical and biological engineering, pharmaceutical industry, medicine, as well as students at different learning levels. For this reason, the first chapter is written mainly for scientists and engineers, as well as other interested specialists, undergraduates, and postgraduates, who are not familiar with the problems of oxidation processes occurring by the mechanism of oxygen atom transfer reactions to substrates. This chapter acquaints the reader with some fundamentals related to the kinetic peculiarities of these reactions, which may be useful for understanding the state-of-the-art in this area of investigation. Present developments in at least two main branches of catalysis are based on achievements in this area of investigation. One of them is catalytic or enzymatic oxidation of organic substrates by the participation of transition metal-oxo compounds in the presence of different oxidants, including dioxygen. The second important area is considered the catalytic or photocatalytic oxidation of water using transition metal-oxo complexes. Both of these branches are of fundamental importance in biology. As a part of the biological evolution of life, Nature carries out these chemical transformations using enzymes, through the reactions of oxo-atom transfer in photochemical formation of oxygen (oxygenic photosynthesis), on the one hand, and its reduction in the respiration processes, on the other hand. In this chapter, the main types of oxidation reactions and the place of oxygen atom transfer reactions in their general classification, from the point of view of the mechanisms, have been discussed. Modern perceptions of the mechanism of oxygen atom transfer reactions in oxidation processes by the participation of transition metal-oxo complexes permit to distinguish at least two main types of reactions. Here, a brief description of these has been presented. The first group of mechanisms involves inner sphere reactions of transition metal-oxo complexes forming an intermediate complex with the substrate with the direct participation of the metallic centers. Then, this intermediate decomposes into an oxygenated product and a reduced form of the initial metal in a complex compound. For the second type of mechanism, named the outer sphere reaction mechanism, it has been considered that the intermediate complex is formed due to the interaction between the oxo-ligand of transition metal complex and the substrate. This chapter addresses the different aspects of the problems of the functionalization of C-H bonds of organic compounds in oxidative catalysis by transition metal-oxo complexes. According to the accepted mechanism, the catalytic cycle involves either the direct transfer of oxygen-atom from the catalyst to the substrate or the hydrogen atom abstraction from the substrate, hydroxylation of metal-ion and subsequent formation of oxygenated products. To perform this catalytic cycle, the reduced metal-ion returns to its initial state being oxidized by another oxidant in the reaction medium. Thermodynamic and kinetic analyses of the catalytic cycles indicate that the major factors determining the reaction mechanism are the energy required to rupture the C-H bonds in oxo-atom transfer reactions and the energy of metal-oxygen bond in re-oxidation of metal. For a successful catalysis, these two energy values must be comparable. These problems are briefly discussed in this chapter. In the last section of the mentioned chapter, the mechanism of oxo-atom transfer reactions has been discussed in light of the phenomenon of multiple spin-state reactivity. It has been exemplified by the reactions of "bare" transition metal-oxo cations $(MO)^+$, where M is a transition metal, with inorganic (H_2) and organic (CH_4) compounds. A great number of theoretical calculations and experimental results indicate that the relationships between the spin states of transition metal-oxo complexes and their reactivity are common for the majority of oxo-atom transfer reactions in the catalysis. In chemical or biological systems, changes in the spin state in transition metal-oxo complexes and, consequently, changes in the reaction pathways permit to explain some of the unusual kinetic features observed in oxo-atom transfer reactions.

The following two chapters of the present volume are scientific reviews devoted to the different aspects of some modern problems of the mechanisms in oxygen atom transfer reactions mainly related to the biological systems. Chapter 2 discusses the mechanisms of oxygen atom transfer reactions related to the bio-inspired activation of dioxygen and its subsequent reactions. The mechanisms of enzymatic oxidation are compared with the schemes of catalytic cycles in oxidation by transition metalorganic complexes as synthetic models of enzymes. In general, this chapter, to some extent, summarizes different catalytic strategies (bio-inspired, biomimetic, synthetic models of enzymes, industrial catalysts) in the activation of dioxygen and its further reactions, including oxygen atom transfer reactions from transition metal complexes to substrates. The bio-inspired activation of dioxygen is exhibited in examples of substrate oxidation by some popular enzymes, such as P450s, monooxygenases, and dioxygenases. Here, the catalytic cycle for P450 is based on the heme-Fe(III) complex, which forms the key intermediate Fe(IV)=O^+ and carries out the hydrogen atom abstraction from RH and further transfer of OH to the substrate. This is a classic example of the oxygen rebound mechanism activating the C-H bonds *via* the radical pathway. A number of other examples demonstrate the widespread importance of oxygen rebound mechanisms in biomimetic chemistry. The analogies and differences of the catalytic cycles of monooxygenases and dioxygenases in bio-inspired oxidation of substrates have been discussed using numerous examples. Here, the discussion is also centered on comparable descriptions of the differences in the enzymatic cycles of dioxygenases with respect to the structural and chemical peculiarities of substrates. For example, according to the proposed schemes, when the pyrrole ring of L-tryptophan is cleaved and two oxygen atoms are inserted into the structure, in the case of tryptophan 2,3-dioxygenase (TDO) and indoleamine 2,3 dioxygenase (IDO), the supplier of four electrons to the oxygen atoms is the same substrate, but in schemes for intradiol ring-cleaving dioxygenases and extradiol dioxygenases, the activation of oxygen requires two electrons from external donor(s) other than the substrate. Special attention has been paid to oxidation systems which are of interest to the chemical and pharmaceutical industries. Among them, the cleavage of C=C bond and stereoselective or asymmetric epoxidation of olefins catalyzed by synthetic transition metalorganic complexes is one of the important areas in modern catalysis. The final section of this chapter covers new catalytic strategies for the activation of dioxygen in oxidation reactions. Among the numerous factors influencing the catalytic activity, the structure of the first coordination sphere of the metal-ions and the surrounding hydrogen bond network is crucial for the successful oxidation of substrates. Apparently, hydrogen bonds play a stabilizing role in the generation of superoxo radicals and promote the cleavage of the O-O bond *via* the formation of metal-oxo moieties. Lewis acids play an analogous role in chemical systems. These perceptions have been demonstrated by the example of vanadium(IV) complexes oxidation schemes. Summarizing the literature data presented in Chapter 2, the authors remark that the creation of efficient industrial catalysts, particularly, in olefin epoxidation, may be achieved using dioxygenase-type enzymes that do not require extra electron suppliers.

Unlike the previous chapter, the third chapter is a review highlighting the peculiarities of oxygen atom transfer reactions from the viewpoint of biomimetic chemistry on the examples of only nickel organometallic complexes. On the occasion of the preparation of this chapter, one of the authors, pr. Sankaralingam, wrote: *In synthetic biomimetic model chemistry, iron and manganese complexes are the most exploited catalysts in the realm of organic transformations reactions. In contrast to a large number of high level and comprehensive reviews reported based on Mn, Fe and Cu oxygen species in various oxidation reactions, relatively less emphasis has been put on nickel oxygen species in oxo-atom transfer reactions. This chapter aims at summarizing the noteworthy attempts in oxo-atom transfer reactions catalyzed by nickel complexes.* In this regard, thorough data are available involving the methods of synthesis, characterization, and revelation of the electronic and geometric structural features of the nickel organometallic complexes, as well as reaction intermediates in the activation of dioxygen and further oxygen atom transfer reactions to substrates. Considerable attention has been paid to the effects of the stereoelectronic properties of the ligand structure on the catalytic efficiency in oxo-atom transfer reactions. Chapter 3 consists of three main paragraphs involving the reactions of oxygen atom transfer and hydrogen atom abstraction catalyzed by nickel organic complexes separately, as well as reactions exhibiting both oxygen atom transfer and hydrogen atom abstraction reactivity jointly. The catalytic role of Ni ions of enzymes, such as glyoxylase I, nickel superoxide dismutase, urease, NiFe hydrogenase, CO dehydrogenase, acetyl-CoA synthase and, methyl-CoM reductase, among others, was the subject of a great number of investigations in biomimetic chemistry. In oxidation processes, involving oxo-atom transfer reactions, as has been shown in this chapter, the active forms of complexes mainly contain Ni(I) and Ni(III), and often Ni(0) and Ni(II) species. In the activation of dioxygen, different nickel oxo, peroxo, superoxo intermediates may be formed, the majority of which are active in oxygen atom transfer or hydrogen abstraction reactions. Of particular interest is the section of Chapter 3 devoted to the discussion of the Ni-complexes exhibiting both the oxygen atom transfer and the hydrogen atom abstraction reactivities. Apparently, these observations are related to the electromeric states of Ni-complexes, (i) Ni^{II}-O$^{\cdot}$ and (ii) Ni^{III}=O, exhibiting different reactivity depending on the nature of substrates (for example, the electrophilicity with PPh$_3$ or CO and nucleophilicity with ArCHO). This review also emphasizes the importance of the ligand architecture in the reactivity of organometallic oxo, dioxo, peroxo superoxo, and hydoperoxo Ni-organic complexes. Usually, their reactivity in oxo-atom transfer reactions correlates with the stereo-electronic properties of the ligands.

The aim of Chapter 4 is to acquaint the reader with the reactions of oxygen atom transfer in the oxidation of organic compounds with dioxygen or other oxidants that occur under visible light or UV irradiation in heterogeneous catalytic systems. Usually, heterogeneous photocatalytic redox reactions occur in multicomponent systems consisting of at least a substrate, oxidant, catalyst, catalyst support, solvent, often also sensitizer. Visible light or UV irradiation may be absorbed by one or more component(s) of the system, which become electronically excited species. Subsequently, they may enter different physical and chemical interactions, transferring energy or electrons to other components involving the catalyst or nominal catalyst. Often, the photochemically generated intermediates, active oxygen species, act as catalysts, for example metal-oxo moieties in transition metal complexes in oxidative catalysis. Two main classes of reactions, namely the photogenerated and catalyzed photolysis, are known in heterogeneous photoredox systems depending on the type of catalyst functionality. Examples of heterogeneous photocatalytic redox reactions, given in this chapter, involving mainly the reactions of organic compounds on TiO_2 or TiO_2-based semiconductor catalysts, demonstrate the predominant role of oxygen atom transfer reactions in the mechanisms of a great number of oxidation or oxidative decomposition processes. Discussing some aspects of the determination of the type of heterogeneous photocatalytic systems, it was concluded that, seemingly, the majority of known heterogeneous photocatalytic reactions on TiO_2, in particular, oxidation through oxygen atom transfer mechanisms, are photoassisted (catalyzed photolysis) processes. Among the oxygen atom transfer agents, transition metal-oxo complexes constitute the main class of compounds widespread in living nature and synthetic chemical systems. Some peculiarities of the photoassisted transfer of oxygen atom in oxidation reactions are discussed in this chapter. Using molybdenum metal-oxo complexes as an example, a significant enhancement of the catalytic activity in oxygen atom transfer on the heterogenization of the homogeneous catalyst was observed. Mo-oxo complexes anchored on TiO_2 with covalent chemical bonds, exhibit improved photocatalytic activity in selective oxidation and oxidative destruction reactions, such as the interaction of O_2 with DDT (dichlorodiphenyltrichloroethane) or other chlorophenyl substituted alkanes, which may not be oxidized at so mild conditions even other strong oxidants. All the examples of photocatalytic reactions mentioned in this chapter also indicate that oxidation occurring by oxygen atom transfer is one of the effective pathways for the creation of new catalytic systems that are economically advantageous and environmentally benign.

Generally, all chapters of this volume introduce not only some fundamentals and state-of-the -art, but also the main directions of development in investigations leading to the revelation of the reaction mechanisms in oxygen atom transfer reactions. For obvious reasons, a separate volume cannot address most of the problems in this field. However, I hope that this volume will be of interest to a wide range of readers, from researchers to students. On the other hand, the discussion of certain problems will apparently give rise to new problems and new interests. This is one of the main aims of creating such a volume.

I am very grateful to academician Jincai Zhao for the foreword for this volume. I would also like to acknowledge the valuable contributions of all authors preparing this volume during a very difficult time for humanity, caused by Covid-19 in the world.

Robert Bakhtchadjian
Institute of Chemical Physics
National Academy of Sciences of the Republic of Armenia
Yerevan,
Republic of Armenia

List of Contributors

Anjana Rajeev
Bioinspired & Biomimetic Inorganic Chemistry Laboratory, Department of Chemistry, National Institute of Technology Calicut, Kozhikode, Kerala-673601, India

Guangjian Liao
School of Chemistry and Chemical Engineering, Huazhong University of Science and Technology, Wuhan 430074, PR China

Guochuan Yin
School of Chemistry and Chemical Engineering, Huazhong University of Science and Technology, Wuhan 430074, PR China

Muniyandi Sankaralingam
Bioinspired & Biomimetic Inorganic Chemistry Laboratory, Department of Chemistry, National Institute of Technology Calicut, Kozhikode, Kerala-673601, India

Robert Bakhtchadjian
Institute of Chemical Physics, National Academy of Sciences of the Republic of Armenia, Yerevan, Republic of Armenia

CHAPTER 1

Introductory Notes on Mechanisms in Oxygen Atom Transfer Reactions of Transition Metal Complexes

Robert Bakhtchadjian[1,*]

[1] *Institute of Chemical Physics, National Academy of Sciences of the Republic of Armenia, Yerevan, Republic of Armenia*

Abstract: Investigations of the mechanisms of oxygen atom transfer reactions of transition metal organometallic complexes are mainly related to their abundance in chemical syntheses and biological oxidation processes. They are important stages in the catalytic and enzymatic oxidation cycles of substrates, as well as in the catalytic oxidation of water.

These brief notes on the mechanisms of oxygen atom transfer reactions involve certain fundamentals (geometric and electronic structures, spin states and reactivity of oxo-complexes), as well as some specific peculiarities of the oxo-atom transfer reactions of transition metal complexes (hydrogen atom abstraction and oxygen rebound mechanisms, intra- and intermolecular types of oxo-atom transfer, multistate reactivity). This chapter introduces readers to the categorization and place of oxo-atom transfer reactions in the classification of catalytic oxidation processes in the context of general problems of the mechanisms in this area. The chapter also provides readers with certain data on the activation of dioxygen and the functionalization of C-H bonds in oxidation processes *via* the oxo-atom transfer reactions of transition metal complexes. The role of the two and multiple spin states reactivity in the mechanisms of these reactions has also been discussed.

This chapter is written mainly for non-specialist readers in this area and serves as a general introduction to the next chapters of this collection of works.

Keywords: Oxygen atom transfer, Catalytic oxidation of water, Multiple spin state reactivity, Oxo-atom transfer in catalytic oxidation, Transition metal-oxo complexes.

** Corresponding author Robert Bakhtchadjian**: Institute of Chemical Physics, National Academy of Sciences of the Republic of Armenia, Yerevan, Armenia; E-mail: robakh@hotmail.com

INTRODUCTION

This chapter of introductory notes on oxygen atom transfer reactions is devoted to the reaction mechanisms of oxidation processes by the participation of transition metal-oxo compounds (transition metal oxides, metalorganic complexes with different ligands, and salts of transition metals). Compared to the next chapters of this volume, it is addressed primarily to readers who are not specialized in this area of investigation. The availability of scientific information to readers unfamiliar with the mechanisms of oxidation reactions requires a preliminary acquaintance with the scientific outlines of general problems. Therefore, this brief chapter, in my opinion, may be useful for researchers, engineers, or students working in neighboring fields, as well as for readers who are first acquainted with the catalytic oxidation reactions by transfer of the oxygen atom of transition metal organometallic complexes.

The revelation of the reaction mechanisms of oxygen atom transfers of transition metals is of pivotal importance for understanding, influencing, and even, controlling the catalytic oxidation of organic or inorganic compounds, including the biological oxidation processes. Oxygen atom transfer reactions are basic stages in the two main domains of oxidation catalysis: (i) oxidative addition of an oxo-atom of transition metals or their complexes with different ligands to organic and inorganic compounds (substrates), and (ii) oxidation of water by the catalysts or enzymes. These catalytic processes are widely used in chemical syntheses on both laboratory and industrial scales and are also basic in understanding the natural biological processes of oxidation occurring through the participation of enzymes. According to the opinion of Gray [1, 2], a pioneer in the area of the electronic structure of metal-oxo complexes, from the point of view of the biological evolution of life on our planet, these two chemical transformations may be considered "top reactions," since one of them is related to the photochemical oxidation of water forming oxygen, and the other is a reduction of oxygen to water during respiration. To carry out these reactions, nature has created complex "machines," enzymes, using transition metal elements. Oxo-atom transfer reactions play an essential role in both basic biochemical transformations. In this regard, the great importance of investigations in this area is evident in the biology, chemical synthesis, medicine, chemical and pharmaceutical industries, *etc*.

Transition Metal-oxo Complexes and their Formation

According to the IUPAC nomenclature, the chemical entities containing one (single) oxygen atom doubly bonded ($=O$) to the atom of another chemical element are termed oxo-compounds [3]. If that chemical element is carbon, the oxo-compounds can be aldehydes, ketones, carboxylic acids, and so on. In another case, if that element is a metal or semimetal (or their ions), the formed chemical entities are metal-oxo (M=O) compounds, for example oxo-molybdenum; oxo-tungsten, *etc.* Here, formally, O^{2-} is a ligand bonded to a metal or metal ion (M^{n+}). Compounds of the d- and f-block transition metals of the periodic table containing oxo ligand(s) and various organic and inorganic ligands, constitute a large class of transition metal-oxo coordination complexes [4, 5].

In oxo-complexes of transition metal elements, an oxygen atom can be bonded to one or more metallic centers. If it is bonded to only one metal atom, the compound is named mononuclear and the oxo-atom is named "terminal-oxo". If the complex is binuclear, the oxo-atom is named double bridged-oxo [4].

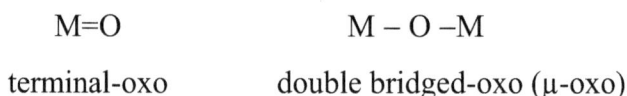

M=O   M – O –M

terminal-oxo   double bridged-oxo (μ-oxo)

Depending on the number of oxo-atoms, the compounds are named mono-,di(bi)-, tri-, polyoxometallic or organometallic complexes. Moreover, the fragments of the chemical structures presented below can be moieties in organometallic complexes of different geometries.

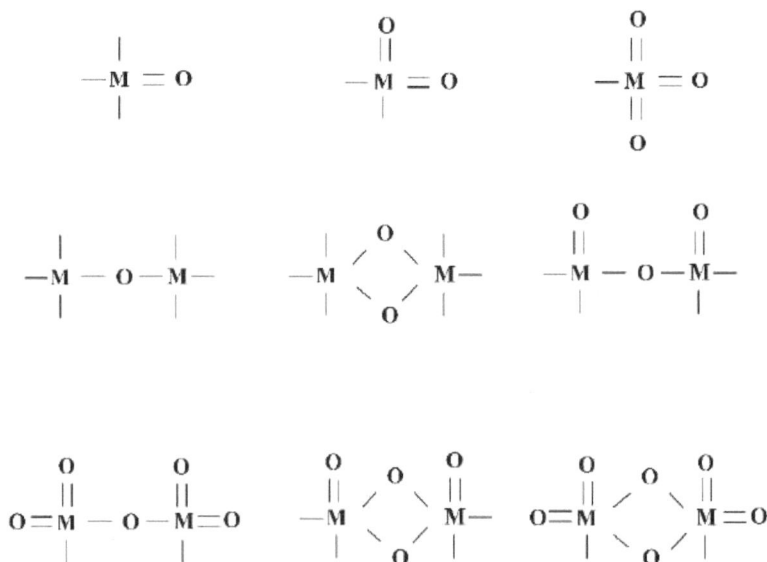

Similarly, organometallic peroxo compounds may also be either mononuclear or polynuclear, on the one hand, and, "end-on" (η^1) or "side-on" (η^2) structures, on the other hand,

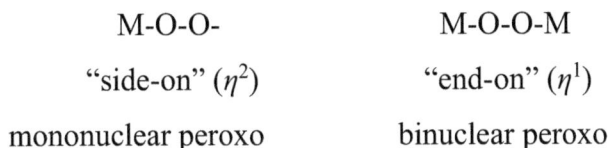

$$M\text{-}O\text{-}O\text{-} \qquad\qquad M\text{-}O\text{-}O\text{-}M$$

"side-on" (η^2)   "end-on" (η^1)

mononuclear peroxo   binuclear peroxo

One of the general methods for the syntheses of transition metal-oxo compounds may be presented by the following reaction scheme (1-3) [6, 7].

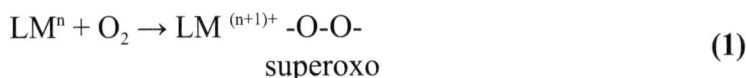

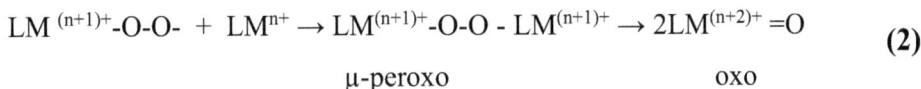

$$LM^n + O_2 \rightarrow LM^{(n+1)+}\text{-}O\text{-}O\text{-}$$
superoxo
$$(1)$$

$$LM^{(n+1)+}\text{-}O\text{-}O\text{-} + LM^{n+} \rightarrow LM^{(n+1)+}\text{-}O\text{-}O\text{-}LM^{(n+1)+} \rightarrow 2LM^{(n+2)+}=O$$
μ-peroxo   oxo
$$(2)$$

$$LM^n + O_2 \rightarrow LM^{n+2}\begin{smallmatrix}O\\|\\O\end{smallmatrix} \rightarrow LM^{n+4}=O$$
$$(3)$$

where L is ligand.

The high-valent metal-oxo M=O may be formed by a direct reaction between the metal ions and O_2 or O_3 in an aqueous solution (for instance, Cr^{3+} with O_2, Fe^{4+} with O_3). In organometallic complexes, the combination of the early transition metal ions with dioxygen and the consequent formation of metal-oxo moieties is favored for two main reasons: first, the central ions of the complexes have high electropositivity, and, second, the partially or entirely empty orbital of metal or metal-ion is available for a π-donation from the ligand L [8].

A great number of high-valent metal-oxo compounds may be obtained by the reaction of oxygen donor molecules with LM^{n+}. The most suitable oxidants may be amine N-oxides (R_3NO), iodosobenzene (C_6H_5IO or analogs of PhIO), sodium periodate ($NaIO_4$), peroxyacids ($RC(O)OOH$), *etc* [9]. For instance, the following reactions (4-6) are an example of the synthesis of nonheme-Fe=O [9].

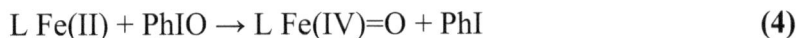

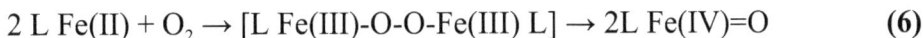

$$L\,Fe(II) + PhIO \rightarrow L\,Fe(IV)=O + PhI \qquad\qquad (4)$$

$$L\,Fe(II) + ROOH \rightarrow L'Fe(III)\text{-}OOR \rightarrow L'Fe(IV)=O + RO^\bullet \qquad (5)$$

$$2\,L\,Fe(II) + O_2 \rightarrow [L\,Fe(III)\text{-}O\text{-}O\text{-}Fe(III)\,L] \rightarrow 2L\,Fe(IV)=O \qquad (6)$$

Application of the Redox Concept in Oxidation Reactions

In general lines, the basic thermodynamic and kinetic principles of heterogeneous or homogeneous-catalytic reactions do not differ from those for enzymatic oxidation processes.

Before examining the peculiarities of oxo-atom transfer reactions in different systems, let us determine their role and place in the classification of the reaction mechanism for catalytic and non-catalytic oxidation processes. These brief notes may be useful for understanding the principal differences between the mechanisms of oxo-atom transfer reactions of different types in oxidation processes.

Note an important peculiarity of the application of the redox concept related to the oxidation reactions in particular cases. The classical concept of oxidation-reduction (redox) reactions in organic chemistry, according to Breslow [10], is not "well defined" and often encounters difficulties in understanding the reaction mechanisms. An obvious example of this is the oxidation reaction of hydrogen with dioxygen in the gas phase. According to the redox concept, here, hydrogen is a donor of electrons and a reductant, whereas oxygen is an acceptor of electrons and an oxidant. However, it is well known that in the gas phase reaction between dioxygen and hydrogen, the main intermediates of the reaction, H, OH, O and HO_2 species, are free radicals, but not ions formed by the direct electron transfers. Therefore, in the given case, the direct application of the redox concept to the reaction

$$2H_2 + O_2 \rightarrow 2H_2O \tag{7}$$

where

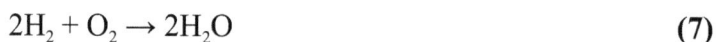

$$2H_2 - 4e \rightarrow 4H^+ \text{ oxidation of hydrogen atoms according to the redox concept}$$

and

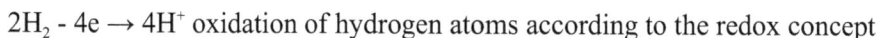

$$O_2 + 4e \rightarrow 2O^{2-} \text{ reduction of oxygen atoms according to the redox concept}$$

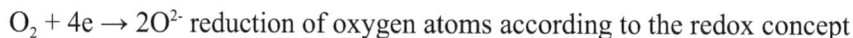

from the point of view of the reaction mechanism, is a formality that has no certain physical meaning.

It is obvious that the insertion of an oxygen atom into a molecule and the formation of an oxygenate is not always oxidation or oxidative addition of oxygen, as in the following reaction (8):

$$CH_2=CH_2 + H_2O \rightarrow CH_3\text{-}CH_2\text{-}OH \qquad\qquad (8)$$

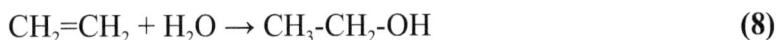

Here, there is no oxidation or reduction, this is a reaction reminiscent of hydrolysis.

This "incomprehension" in the applications of the redox concept may be overcome if the type of the reaction mechanism and its steps, including the steps of electron transfers, were precisely determined. In general, the redox concept is useful and directly applicable to a large number of reactions, including many oxidation reactions that occur through oxo-atom transfer mechanisms. Note that, in general, oxo-atom transfer reactions have either concerted or stepwise mechanisms. They occur by radical (or so-called "rebound") mechanisms when metal-oxo species exhibit properties inherent to oxyl radicals [5].

Types of Reaction Mechanisms in Oxidation Processes and the Place of the Oxo-atom Transfer Reactions within their Classification

The general classification of chemical reactions and their mechanisms may be based on various physical and chemical properties of the systems under study [11]. From the point of view of the phase states (gas, liquid, solid) of the initial reactants and products, all chemical reactions can be divided into heterogeneous, homogeneous, and bimodal (complex reactions consisting of both heterogeneous and homogeneous constituents) [12]. If all reaction components of the system are in the same phase, the reaction is homogeneous, and if they are in different phases the reaction is heterogeneous. In turn, the overall reaction may be either catalytic or non-catalytic. According to Sheldon *et al.* [13 - 15], from the point of view of the reaction mechanisms, the oxidation reactions (heterogeneous or homogeneous, as well as catalytic, non-catalytic, or autocatalytic) may be divided into three main types. Oxidation of a very great number of inorganic and organic compounds occurs by free radical mechanisms (**1**), which are essentially different from two other types of catalytic oxidation reactions: metal or semimetal (ions) coordinated substrate oxidation (**2**), and oxygen atom transfer reactions (**3**). The latter also includes the mechanisms of oxo-atom transfer reactions of transition metals and their organometallic complexes. Both the second and third types of oxidation mechanisms related to the participation of metals or metal ions interacting with the substrates were named heterolytic by Sheldon [14], in comparison with the chain-radical reactions often named homolytic (taking into consideration the formation of radicals by the homolytic cleavage of chemical bonds) [15].

1. In *oxidation reactions occurring by radical and chain mechanisms*, the active intermediates are free radicals, the reactions of which in the fluid phases are chain processes [16]. A chain reaction can be initiated by the generation of free radicals in the reaction medium through the application of catalysts (initiators), heating, light irradiation, or electromagnetic waves of different lengths. One of the peculiarities of the chain-radical oxidation reactions, particularly, with dioxygen, may be considered the feasibility of the autocatalytic progression of reaction, due to the degenerate branching of chains *via* the radical decomposition of main intermediates. Usually, in the oxidation of hydrocarbons or oxygen-containing organic compounds with dioxygen, hydrogen peroxide or organic peroxides are intermediates responsible for the degenerate (partial) branching of chains.

In a homogeneous reaction system, the oxidation can be accelerated by the addition of variable-valence metal ions (including transition metals or their complexes with organic or inorganic ligands). They accelerate the decomposition of peroxy compounds by the formation of free radicals, which partially participate in the propagation and branching of radical chains. Usually, the following reaction pathway may be taken into consideration in a homogeneous reaction environment in the presence of metal ions [14]:

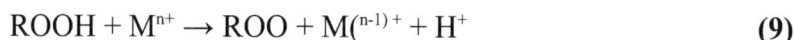

$$ROOH + M^{n+} \rightarrow ROO + M^{(n-1)+} + H^+ \qquad (9)$$

$$ROOH + M^{(n-1)+} \rightarrow RO + OH^- + M^{n+} \qquad (10)$$

In particular case, if the peroxide compound is H_2O_2 and M is Fe^{2+} ion in an aqueous solution, which is known as the Fenton reagent [17], apparently, the formation of radicals occurs according to the reaction mechanism first proposed by Willstatter, Haber and Weis in the early1930s [18, 19]:

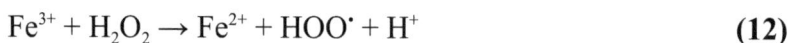

$$Fe^{2+} + H_2O_2 \rightarrow Fe^{3+} + HO^{\bullet} + OH^- \qquad (11)$$

$$Fe^{3+} + H_2O_2 \rightarrow Fe^{2+} + HOO^{\bullet} + H^+ \qquad (12)$$

$$2\,H_2O_2 \rightarrow HO^{\bullet} + HOO^{\bullet} + H_2O \qquad (13)$$

Note that this categorization of oxidation reactions given by Sheldon [13, 15] is mainly based on data obtained in homogeneous liquid phase chain reactions of oxidation. However, it may be extended to heterogeneous systems involving oxidation reactions occurring through radical-like intermediates because, in this case, there are profound similarities between the mechanisms of heterogeneous and homogeneous oxidation reactions [12]. Examples of the formation of radicals in the decomposition of hydrogen peroxide and organic peroxy compounds by the heterogeneous pathway on the surfaces of solid substances, and their further participation in the gas phase chain-oxidation reactions have been revealed by Nalbandyan and Vardanyan in the 1980s [20].

One example of homolytic oxidation by the radical pathway is the reaction of arylalkane with O_2 in the presence of Co(III) acetate, the mechanism of which is represented in a simplified form by Sheldon and Koshi [21] as the following sequence of reactions:

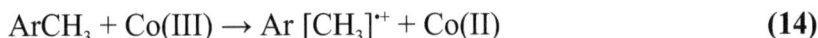

$$ArCH_3 + Co(III) \rightarrow Ar[CH_3]^{\cdot+} + Co(II) \qquad (14)$$

$$Ar[CH_3]^{\cdot+} \rightarrow Ar[CH_2]^{\cdot} + H^+ \qquad (15)$$

$$Ar[CH_2]^{\cdot} + O_2 \rightarrow Ar\,CH_2OO^{\cdot} \qquad (16)$$

$$Ar\,CH_2OO^{\cdot} + Co(II) \rightarrow Ar\,CH_2OOCo(III) \qquad (17)$$

$$Ar\,CH_2OOCo(III) \rightarrow ArCHO + HOCo(III) \qquad (18)$$

In this scheme, reaction (17) can be represented as a sequence of the following reactions (17a-17c).

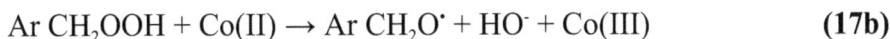

$$Ar\,CH_2OO^{\cdot} + ArCH_3 \rightarrow Ar\,CH_2OOH + Ar[CH_2]^{\cdot} \qquad (17a)$$

$$Ar\,CH_2OOH + Co(II) \rightarrow Ar\,CH_2O^{\cdot} + HO^- + Co(III) \qquad (17b)$$

$$Ar\,CH_2OO^{\cdot} + Co(III) \rightarrow Ar\,CH_2OO\,Co(III) \qquad (17c)$$

Why are the radical mechanisms so widespread in oxidation chemistry? Since most organic compounds under normal conditions are in a singlet electronic state and dioxygen is in a triplet electronic state, the formation of oxygenates by the direct combination of organic compounds with dioxygen in a concerted reaction is spin-forbidden by the rules of quantum mechanics (see the section "Spin state and reactivity"). One way to circumvent this restriction is the stepwise occurrence of reaction as a chain-radical process. Note that the chain-radical reaction comprises at least three main stages: initiation, propagation and termination of chains. In a reaction medium, the homogeneous and heterogeneous catalysts play different roles participating in one or more stages of the chain-radical process [12].

Some authors use the so-called oxidizability parameter $k_p/(2k_t)^{1/2}$ (where k_p and k_t are the reaction rate constants of the chain propagation and termination stages, respectively), when comparing the reactivity of different substrates [22]. This is a kinetic parameter that depends on the reaction conditions. Note also that this approach is applicable, if the reaction occurs only according to the "purely" chain-radical mechanism, mainly during the liquid phase oxidation.

2. The second type of mechanism comprises the oxidation reactions of organic compounds by the participation of metal ions or metal complexes. Their characteristic peculiarity is the formation of an intermediate coordination complex with the substrate, which in turn can be decomposed to an oxidized product and a reduced metal. The process may become catalytic if the reduced metal or metal ion undergoes oxidation under reaction conditions with another ("second") oxidant agent. A classic example is the Wacker process (19-21) [23, 24].

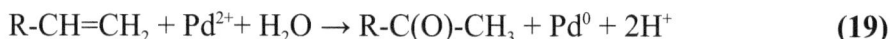

$$R\text{-}CH=CH_2 + Pd^{2+} + H_2O \rightarrow R\text{-}C(O)\text{-}CH_3 + Pd^0 + 2H^+ \tag{19}$$

In the Wacker process, Pd can be re-oxidized indirectly, in the presence of $CuCl_2$ and O_2.

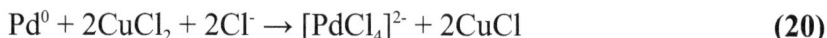

$$Pd^0 + 2CuCl_2 + 2Cl^- \rightarrow [PdCl_4]^{2-} + 2CuCl \tag{20}$$

$$2CuCl + 1/2O_2 + 2HCl \rightarrow 2CuCl_2 + H_2O \tag{21}$$

The following scheme presents the proposed mechanism of an important reaction (19) of the coordination of ethylene with Pd ions and the subsequent formation of acetaldehyde [24, p.65 - 67].

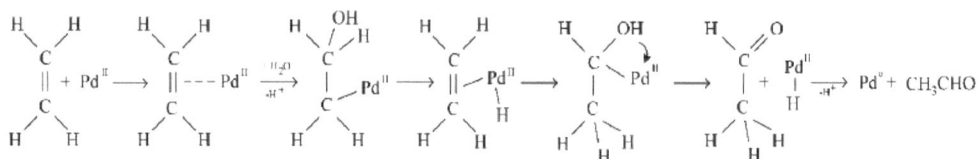

According to this mechanism, the process begins with the coordination of $CH_2=CH_2$ with Pd^{2+} ions, partially replacing Cl^- ions in the coordination sphere of $[PdCl_4]^{2-}$ (note that $PdCl_2$ is insoluble in water, while $[PdCl_4]^{2-}$ is soluble). OH^- and H_2O can also enter the coordination sphere of the central Pd(II) ion. Apparently, the nucleophilic attack of water on ethylene leads to the formation of an intermediate, hydroxylethylpalladium, which is accompanied by the displacement of H^+, although this intermediate has not been experimentally identified. Then, the formation of vinyl alcohol can occur *via* the β-hydride elimination. Finally, the abstraction of a proton with the formation of unstable palladium hydride results in the formation of acetaldehyde. Here, the essential role of the coordination of the metal ion with the substrate in the oxidation process is obvious [24, p.65].

The palladium (II) catalyst is also successfully used in other reactions, for example, in the oxidative dehydrogenation of alcohols. The reaction mechanism involves the coordination of the substrate by the complex metal ion, the formation of an intermediate product, and further dehydrogenation, which leads to the formation of aldehydes or ketones [25].

3. The third type of reaction mechanism in oxidation processes, which is of particular interest from the point of view of this work, is *catalytic oxygen transfer* (this term was also proposed by Sheldon [14]). This includes oxidation reactions with strong oxidants in the presence of metals, metal oxides, or their coordination complexes with different ligands.

The first simplest examples of catalytic oxygen transfer reactions may be considered the oxidation reactions of alkenes with a mixture of metal oxides (OsO_4, MoO_3, V_2O_5, CrO_3) and hydrogen peroxide in *tert*-butanol solution, since the 1930s referred to as the Milas reagents [26].

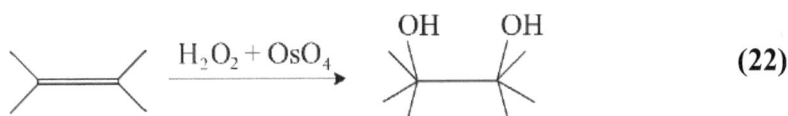

$$\text{(22)}$$

In general, oxidants are metal-oxo or metal-peroxo compounds that interact with the substrates under homogeneous or heterogeneous conditions.

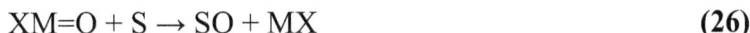

$$MX + H_2O_2 \ (RO_2H) \rightarrow M\text{-}O\text{-}O\text{-}H \ (MOOR) + HX \tag{23}$$

$$MX + H_2O_2 \ (RO_2H) \rightarrow XM{=}O + H_2O \ (ROH) \tag{24}$$

$$M\text{-}O\text{-}O\text{-}R + S \rightarrow SO + MOR \tag{25}$$

$$XM{=}O + S \rightarrow SO + MX \tag{26}$$

In this scheme, MX is a transition metal compound that forms metal-oxo or peroxo moieties with H_2O_2. In the reaction with an alkene, metal-oxo or metal-peroxo compounds transfer an O-atom forming an epoxide compound. It was then transformed into vicinal diol, apparently by hydrolysis. Thus, this type of reaction also includes oxo- or peroxo-type of oxygen atoms transfer to the substrate, catalyzed by metal ions (metal oxides, organometallic complexes).

In heterogeneous catalysis, this type of mechanism is similar to the Mars-van-Krevelen mechanism [12, 27] which may also be included in the Sheldon classification of oxidation mechanisms. Obviously, the determination of the type of oxidation mechanism is more complicated in biological systems, in a complex heterogeneous-homogeneous environment, where various factors play an important role in the formation and reactions of intermediates.

It should be noted that the division of the suggested mechanisms of oxidation reactions into the above-mentioned types, is somewhat conditional. In the following sections, it will be shown the feasibility of the appearance of the so-called multistate reactivity that is related to the existence of multiple spin states of the metal-oxo moieties. For instance, FeO^+ complexes in the functionalization of C-H bonds of hydrocarbons in the gas phase can exhibit two-state reactivity. The spin states of the reactants can be changed during the reaction. As a result of the changes in the spin state, the reaction mechanism also changes, passing from one potential energy surface to another, energetically more favorable one. The FeO^+ species in the reaction with methane in the gas phase, subsequently exhibit either biradical properties, and the reaction occurs as a stepwise radical process, or the properties of ordinary nonradical species, and the reaction occurs as a concerted process. Thus, in certain cases, there is no "purely radical" or "purely non-radical" reaction in oxygen atom transfer processes.

Finally, let us mention another peculiarity of the mechanism of oxidation reactions. In the study of oxidation reactions, the overall process cannot always be divided into "purely" homogeneous or "purely" heterogeneous reactions occurring by free radical, electron transfer or other mechanisms. In a great number of cases, the oxidation reactions exhibit a bimodal character involving both homogeneous and heterogeneous constituents, each of which may be described using different reaction mechanisms [12].

The Role of Oxo-atom Transfer Reaction in Catalytic Oxidation Processes

As mentioned in the introduction, the two main areas of catalytic oxidation processes involve oxo-atom transfer reactions of transition metal complexes.

I. Catalytic or enzymatic oxidation of organic substrates *via* transition metal-oxo or peroxo compounds in the presence of oxidant(s) other than the oxo-atom of metal (ion) in the reaction medium [4 - 8],

II. Catalytic oxidation of water by the participation of transition metal-oxo compounds.

I. In the case of heterogeneous or homogeneous catalytic oxidation of substrates by metal-oxo compounds, the following catalytic cycle takes place,

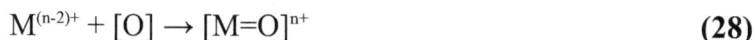

$$[M=O]^{n+} + S \rightarrow M^{(n-2)+} + SO \qquad (27)$$

$$M^{(n-2)+} + [O] \rightarrow [M=O]^{n+} \qquad (28)$$

where S is an organic substrate.

The sources of oxygen atoms [O] can be O_2, H_2O_2, DMSO, HJO, ROOH, lattice oxygen, *etc*. Here, they may be referred to as the second oxidant with respect to the primary oxidant $[M=O]^{n+}$. A number of oxidation processes in the chemical and pharmaceutical industries are based on this scheme [28]. Some important details of this catalytic cycle are provided in Chapter 4.

II. The summary reaction of water oxidation using transition metal-oxo compounds as catalysts is as follows [29 - 31]

$$2[M=O]^{n+} + 2H_2O \rightarrow 2M^{(n-2)+} + O_2 + 4H^+ + 4e^- \tag{29}$$

From the point of view of practical use, the splitting of water is a source of hydrogen, usable in different areas, including energy production.

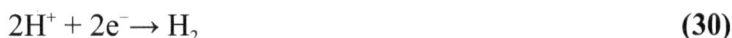

$$2H^+ + 2e^- \rightarrow H_2 \tag{30}$$

The best catalysts for this reaction, presently, may be considered Ru, Co and Ir organometallic complexes containing pyridine ligands [29]. There are catalytic, photocatalytic and electrocatalytic pathways of water splitting [29, 30]. For instance, in homogeneous catalysis, $[Ru^{III}(bipy)_2(H_2O)_2O]^{4+}$ complex can oxidize H_2O [29 (p.42), 31]. In heterogeneous catalysis a number of metal oxides, such as Ru, Ir, Fe, Co, Mn, are good catalysts for water oxidation [29]. Among them, IrO_2 is known to be a catalyst with higher turnover numbers [29].

Electronic Structure and Nature of the Metal-oxygen Chemical Bonds in Metal-oxo Complexes

The chemical properties of transition metal-oxo complexes in oxidation reactions are related to the specificity of the chemical bonding between the metal ion and the oxygen atom in the ligand- or crystal-field. As it has been mentioned above, formally, an oxo-complex may be regarded as a compound in which oxy dianion O^{2-} is bonded with the metal ion $M^{(n+2)+}$. As a rule, O^{2-} ions are nucleophile species. However, many experimental results indicate that oxygen atoms in oxo-atom transfer reactions exhibit either electrophilic or nucleophilic reactivity [32 - 34]. The explanations of these observations are related to the concept of partial or full electron transfer from the metal (metal-ion) to oxygen by the formation of different resonance structures of the metal-oxo moieties under certain conditions [32, 33]. According to Yamaguchi *et al.* [34], the transfer of the following resonance forms of oxo-complexes from one to another may be observed depending on the nature of the metal, ligand and reaction environment:

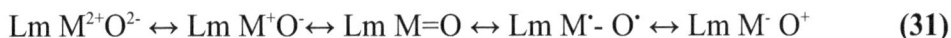

$$Lm\ M^{2+}O^{2-} \leftrightarrow Lm\ M^+O^- \leftrightarrow Lm\ M=O \leftrightarrow Lm\ M^{\cdot}-O^{\cdot} \leftrightarrow Lm\ M^-\ O^+ \tag{31}$$

Lm is ligand.

For a metal ion M^{n+}, these variable resonance forms of the metal-oxo species (31) can be rewritten as

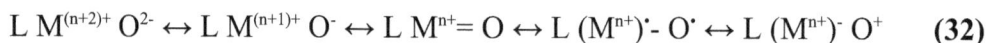

$$L \, M^{(n+2)+} \, O^{2-} \leftrightarrow L \, M^{(n+1)+} \, O^{\cdot-} \leftrightarrow L \, M^{n+} = O \leftrightarrow L \, (M^{n+})^{\cdot} - O^{\cdot} \leftrightarrow L \, (M^{n+})^{-} \, O^{+} \quad \textbf{(32)}$$

L is a ligand other than the oxo-atom, and M is a transition metal. As an example of the mentioned resonance forms, taken from this row, let us consider $LM^{\cdot}$- $O^{\cdot}$ complexes containing metal-oxyl moieties. Recently, their formation, electronic structure, properties and reactivity, particularly, their possible role in the oxidation reactions, have been presented in detail in a review article by Shimoyama and Kojima [33]. According to these authors, $-O^{\cdot}$ is a radical-ligand that can be formed from a metal-oxo compound in the $(n+1)+$ oxidation state of transition metal and represented as:

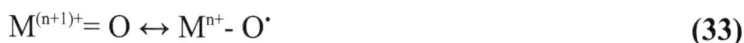

$$M^{(n+1)+} = O \leftrightarrow M^{n+} - O^{\cdot} \quad \textbf{(33)}$$

Examples include metal-oxyls Zn(II)-$O^{\cdot}$; Fe(IV)-$O^{\cdot}$ and Ru(II)-$O^{\cdot}$. Moreover, the appearance of the triplet biradical form of an oxo-atom reminding $O(^{3}P)$ is postulated in a number of investigations [33, and within].

Transition metals or their cations, with some exceptions, have partially filled d- or f-subshells (elements of the d- and f-blocks of the Periodic Table). For example, divalent cations of transition metal elements have the following electronic configurations of the d-atomic orbital (in this row Zn is considered as a transition metal) [35a]:

Sc^{2+}	Ti^{2+}	V^{2+}	Cr^{2+}	Mn^{2+}	Fe^{2+}	Co^{2+}	Ni^{2+}	Cu^{2+}	Zn^{2+}
$3d^{1}$	$3d^{2}$	$3d^{3}$	$3d^{4}$	$3d^{5}$	$3d^{6}$	$3d^{7}$	$3d^{8}$	$3d^{9}$	$3d^{10}$

For example, Fe^{2+} ion has an electronic configuration d^{6} and the following spin state.

$$(\,\uparrow\downarrow \quad \uparrow \quad \uparrow \quad \uparrow \quad \uparrow\,)$$

In the octahedral crystal field (as well as in the ligand field) the five degenerate atomic orbitals of a d-metal or metal-ion are split into two energetically different groups of orbitals: e_g (high energy): $d_{x^2-y^2}$, d_{z^2} ; and t_{2g} (low energy): d_{xy}, d_{xz}, d_{yz}. (Fig. **1**) [35 b.36].

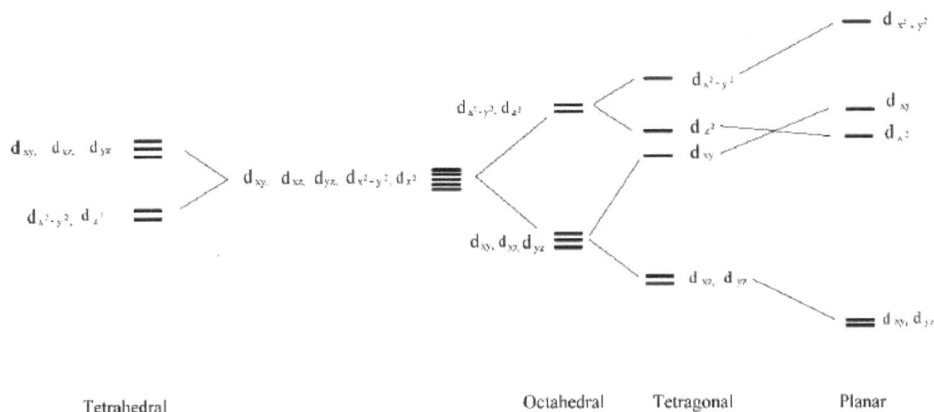

Fig. (1). Energy diagram of the splitting of d orbitals of transition metals in the crystal field [35 b].

Two important parameters determine the probability of the appearance of these energetic states. One is the crystal field splitting energy (Δ) and the second is the spin pairing energy (P). If Δ>P (strong-field), the electrons tend to occupy lower energy orbitals. If Δ<P, the electrons tend to occupy the orbitals of higher energy levels [35, 36].

Two energy diagrams of the molecular orbitals of different metal-oxo moieties are presented in Fig. (2). It is obvious that in complexes with the electronic configuration of metal d^0, the chemical bond between the metal and oxygen is double and consists of σ and π molecular orbitals (Fig. 2a). In the case of d^2-metal, the triple bond between the metal and oxygen dianion consisting of σ, π and δ molecular orbitals for octahedral complexes is shown Fig. (2b). The sequence of molecular orbitals with respect to their energies is the following: σ < π < $d_{x2-y2/xy,xz}$ < π^* < σ^*... Then, when n>2 in the electronic configuration of the metal (d^n), the electrons will also fill the next non-bonding and antibonding π^* and σ^* molecular orbitals [36]. This leads to a decrease in the bond order and a gradual destabilization of the complex. Therefore, it is not surprising that transition metal complexes with electronic configurations d^0, d^1, d^2, such as terminal oxo complexes W(VI, V, IV), Mo(VI, V, IV), V(V, IV), are thermodynamically stable, while in the case of the electronic configurations d^3, d^4 ... they are less stable and more reactive. In general, a comparison of the chemical properties of the early (elements of 3-7 groups) and later (elements of 8-12 groups) transition metal oxides show that, for octahedral complexes, the corresponding metal-oxo bonds are stronger for most of the early transition metal compounds than for the later transition metal compounds containing an oxo-ligand weakly bonded to the metal. In other words, the oxo-complexes of early transition metals are thermodynamically and kinetically stable. The difference between the metal-oxo bond energies of these two groups of elements is about twice (Table **1**) [37].

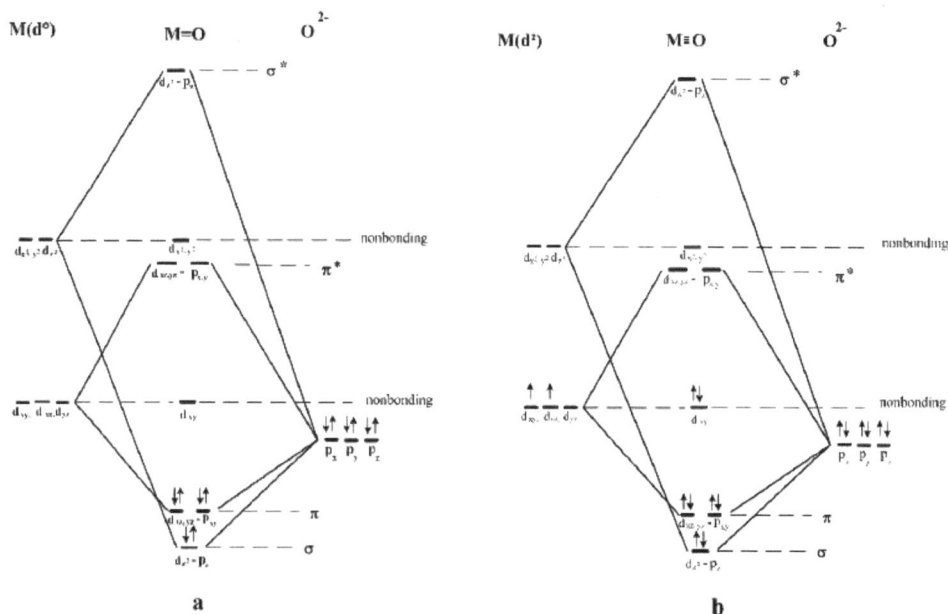

Fig. (2). Energetic diagrams of the metal-oxo moieties of the complexes, where metal ions have: (**a**) d^0 electronic configuration and form double chemical bond with oxygen dianion; (**b**) d^2 electronic configuration and form triple chemical bond with oxygen dianion, in octahedral crystal field. (Adapted from Ref [36].).

Table 1. First Row Transition Metal-oxo Bond Strengths (kcal/mol)*.

Metal	$D_0(M^+ - O)$	$D_0(M-O)$	Metal	$D_0(M^+-O)$	$D_0(M-O)$
Sc	159 ± 7	161 ± 3	Mn	57 ± 3	85 ± 4
Ti	161 ± 5	158 ± 2	Fe	69 ± 3	93 ± 3
V	131 ± 5	146 ± 4	Co	64 ± 3	87 ± 4
Cr	85 ± 1.3	110 ± 2	Ni	45 ± 3	89 ± 5

D_0 is energy of chemical bond between the metal (M) or metal ion (M+) and oxygen
*Taken from the work: Kang H., Beauchamp J. L. J. Am. Chem. Soc. 1986,
v.108, p. 5663, and references therein, they were represebted also in Ref. [37].

Thus, one may state that, in general, the electrophilic properties of metal-oxo species are related to the relatively low negative charge of oxygen, compared to the case of a simple double bond. The weakness of the chemical bonding of metal-oxo often leads to the formation of atomic oxygen, a biradical in the triplet electronic state $\cdot O \cdot$ (^3P), as it has already been mentioned above. The properties of metal-oxo complex compounds depend mainly on the formal oxidation number and charge of the metal-ion [3, 6, 7]. For the first time the nature of the triple chemical bonding between the early transition metals V^{2+} and Mo^{3+} and oxygen was established by Ballhausen and Gray in the 1960s [1, 2, 35a, 38]. This finding permitted him to give an explanation to a phenomenon named "oxo-wall." It is an empirical observation showing that the terminal metal-oxo complexes of elements from 9 to11 groups are very rare, and they have square-pyramidal, trigonal-pyramidal, and square-planar, but not octahedral, geometries. The observed phenomenon divides the transition metal elements between 8 and 9 groups as an "oxo-wall" (Fig. **3**).

Sc	Ti	V	Cr	Mn	Fe		Co	Ni	Cu	Zn
Y	Zr	Nb	Mo	Tc	Ru		Rh	Pd	Ag	Cd
Lu	Hf	Ta	W	Re	Os		Ir	Pt	Ag	Hg

Fig. (3). Oxo-wall.

In other words, the existence of stable complexes with terminal oxo ligands is not possible in the field of tetragonal ligand for elements with more than 5 electrons in the d orbital. However, oxo complexes beyond the oxo-wall, as intermediates are often postulated to be involved in the mechanisms of oxidation reactions [33].

General Types of Reactions of Metal-oxo Complexes

In general, the chemical reactions typical of metal-oxo compounds can be divided into at least three main groups: **oxo-atom transfer**, **hydrogen atom abstraction,** and **oligomerization** (so-called olation or condensation) reactions [39]. In the present work, we focus mainly on the oxo-atom transfer reactions, which are also related to the hydrogen atom abstraction reactions.

Mechanisms of Oxygen Transfer Reactions of Metal-oxo Complexes

According to Holm [4], oxygen atom transfer reactions have been known since 1912 based on the work of Hofmann, who investigated the oxidation of maleic acid with OsO_4 and the *in situ* oxidation of OsO_2 by aqueous chlorate to initial metal oxide.

The two main groups of oxygen transfer reactions are known as intra- and intermolecular transfers of oxo-atoms in oxidation [40].

Oxygenation of arene C–H and C–F bonds into C-O by tetranuclear iron complexes $[LFe_3(PhPz)_3OFe]^{2+}$ and $[LFe_3(FArPz)_3OFe]^{2+}$ in interaction with oxygen atom transfer reagents (PhIO) is an example of an intramolecular-type reaction, apparently, occurring through the formation of the terminal iron-oxo intermediate species and further transfer of the oxygen atom [40, 41].

The second group of oxo-atom transfer reactions, intermolecular reactions, is more widespread than the intramolecular reactions [42].

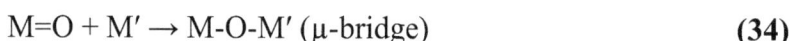

$$M=O + M' \rightarrow M\text{-}O\text{-}M' \ (\mu\text{-bridge}) \tag{34}$$

where M=O is oxygen donor and M' is acceptor. This type of oxo-atom transfer reaction (34), according to Holm [4], is incomplete, in comparison with the following reaction (35) named complete transfer of oxo-atoms

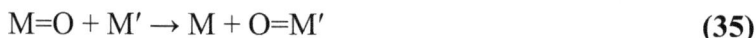

$$M=O + M' \rightarrow M + O=M' \tag{35}$$

Reaction (34) is typically two-electrons transfer redox process, whereas reaction (35) involves a net one-electron transfer [42]. Note that reactions (34) and (35) may more often be consecutive steps of the oxidation process.

The reactivity pattern of the intermolecular oxo-atom transfer is especially rich for the compounds containing high-valent transition metal oxides [4 - 7]. For instance, $[Fe(IV)=O]^{2+}$ complexes (as complexes of $[(N_4Py) Fe(IV)O]^{2+})$ are capable of oxidizing phosphines, organic sulfides, alkenes (cycloalkenes), *cis*-stilbene, and even some saturated hydrocarbons as cyclohexane [43].

The mechanisms of oxygen atom transfer from metal-oxo complexes in the oxidation of organic or inorganic compounds and substrates in chemical and biological systems may be conditionally divided into two main types [44]:

1. The chemical compound or substrate interacts directly with the metal center to form an intermediate complex, the decomposition of which provides an oxygenate product and a reduced form of the metal in an organometallic complex. Thus, the primary reaction involves the inner sphere of the metal-oxo complex. Obviously, depending on the nature of the interaction of the substrate with the coordinated metal, the reaction mechanisms may be different. Examples of inner-sphere interactions include σ-bond metathesis, oxidative addition of a substrate or electrophilic substitution of a substrate to the metal center, and others. In electrophilic substitution reactions, a cation coordinated three-center intermediate may be formed.

$$L M{=}O + R-H \longrightarrow \left[\begin{array}{c} H \cdots\cdots R \\ \diagdown \diagup \\ LM^+ \end{array} + O^- \right] \longrightarrow LM + ROH \quad (36)$$

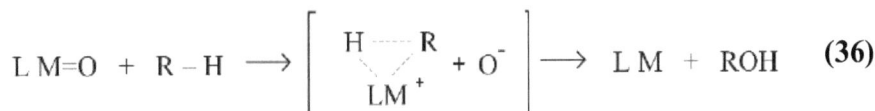

2. Oxygen atom transfer takes place during the interaction of the substrate with the oxo-ligand of transition metal complex. Therefore, this may be referred to as an outer-sphere oxo-atom transfer reaction. There are two different reaction mechanisms related to the oxo-atom transfer: concerted (37) and stepwise (radical or oxygen rebound mechanism, 38) occurrence of the oxygen atom transfer.

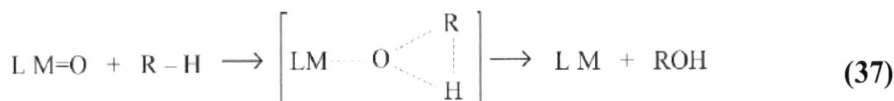

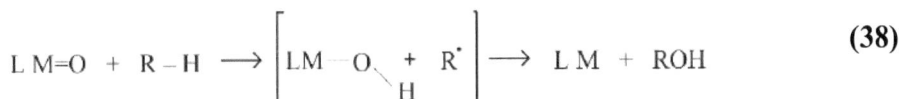

$$L M{=}O + R-H \longrightarrow \left[LM \cdots\cdots O \begin{array}{c} R \\ \diagup \\ H \end{array} \right] \longrightarrow LM + ROH \quad (37)$$

$$L M{=}O + R-H \longrightarrow \left[LM \cdots O \begin{array}{c} + R^\bullet \\ \diagdown \\ H \end{array} \right] \longrightarrow LM + ROH \quad (38)$$

Note that radical reactions are usually not very selective processes, and different byproducts (such as LM-OR, RHO and others) can be expected to be formed.

In the case of the catalytic oxidation by the participation of high-valent transition metal organometallic complexes, the summary reaction corresponds to the scheme [7, 8].

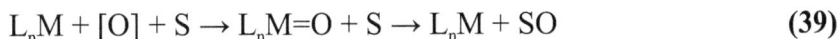

$$L_nM + [O] + S \rightarrow L_nM{=}O + S \rightarrow L_nM + SO \quad (39)$$

L_n is ligand S substrate and [O] is O_2, H_2O_2, DMSO, HJO, ROOH, *etc.* Among these oxidants, dioxygen is the most available and cheapest reagent. In the case of dioxygen, the oxidation has some peculiarities observed for high-valent transition metal-oxo and dioxo (Mo, W, Ru, *etc.*) organometallic complexes used as catalysts or intermediates. Let us consider one of these. Experimental data indicate a quantitative compliance to the stoichiometry of the reaction (40), occurring through oxo-atom transfer to the substrate in the oxidation of a number of chlorinated aryl alkanes, using Mo-dioxo organometallic complexes (for example, dioxo-molybdenum(VI)-dichloro[4,4'-dicarboxylato-2,2'-bipyridine], anchored on TiO_2 support) [45].

$$\begin{array}{c} O \\ \parallel \\ LM{=}O \end{array} + S \longrightarrow LM{=}O + SO \quad (40)$$

In the presence of dioxygen in the reaction medium, oxidation of the substrate becomes catalytic. A significant self-acceleration of the reaction and an almost two-fold increase in the turnover number and the yield of oxygenated products were observed. One of the possible explanations for these observations may be the consideration of the following reactions (41-42) of the formation of intermediate oxo-peroxo moieties, in the mechanism of the overall process:

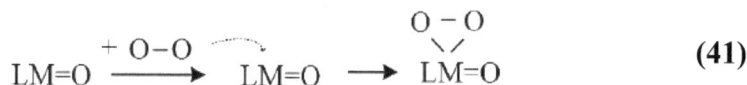

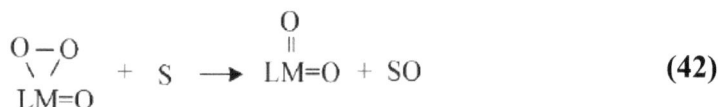

$$\text{LM=O} \xrightarrow{\; + \text{O–O} \;} \text{LM=O} \longrightarrow \underset{\text{LM=O}}{\overset{\text{O – O}}{\diagdown\diagup}} \qquad (41)$$

$$\underset{\text{LM=O}}{\overset{\text{O – O}}{\diagdown\diagup}} + \text{S} \longrightarrow \overset{\text{O}}{\underset{\text{LM=O}}{\|}} + \text{SO} \qquad (42)$$

In general, there is a good agreement between the experimental data and the stoichiometry of reactions 40-42. Thus, this reaction is an example of oxidation when both oxygen atoms of dioxygen are fixed and used in the catalytic oxygenation of substrates without by products.

Functionalization of the C-H Bond *via* Abstraction of a Hydrogen Atom. Oxygen Rebound Mechanism

One of the central problems in the chemistry of transition metal-oxo complexes is the revelation of the mechanism of C-H bond functionalization in substrates under conditions, providing high selectivity that is available only in certain enzymatic processes. Hydrogen atom abstraction from the substrate is a reaction of fundamental importance in transition metal-oxo chemistry. There are two possible mechanisms for hydrogen abstraction from the substrate [46 - 48]. First, the reaction involves electron transfer from the substrate to the metal center of the metal-oxo complex and a combination of proton to an oxygen atom, also known as proton-coupled electron transfer (PCET) [46]. The second mechanism is a one-step and direct transfer of the hydrogen atom from the substrate to the metal-oxo. Usually, in both cases, the overall reaction leads to the formation of an oxygenated substrate *via* the elimination of OH species and oxygen rebound to form ROH. For example, the high-valent transition metal-oxo porphyrinoid complexes, such as $-Cr^V(O)$, $-Mn^V(O)$ and $-Fe^V(O)$, exhibit high reactivity in hydrogen atom transfer reactions from substrates [47]. Studies on the thermodynamic aspects of hydrogen atom abstraction reactions show that synchronicity or asynchronicity in the case of H^+/e^- transfer is a determining factor in a two-step reaction [48]. These two types of mechanisms may be differentiated by applying the method of the primary kinetic isotope effect in investigations [46, 49].

The so-called "oxygen rebound" reaction mechanism [50] has been suggested in many enzymatic oxidation processes *via* the abstraction of hydrogen from the substrate and the formation of radical R.

$$LM^{n+}=O + RH \rightarrow LM^{(n-1)+}- OH + R^\bullet \tag{43}$$

In this mechanism, the primary reaction of H-abstraction from C-H occurs by the formation of the so-called "intermediate cage" [50]. If the central ion is Fe^{n+}, the formation of the following "caged" intermediate is suggested,

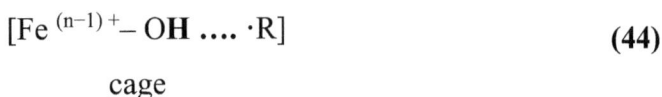

$$[Fe^{(n-1)+}- OH \ .... \ \cdot R] \tag{44}$$
$$\text{cage}$$

This intermediate can undergo different transformations. One of them is oxygen rebound to form the product

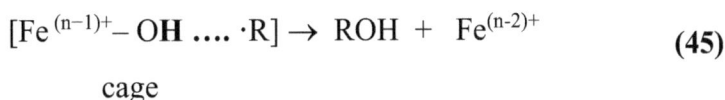

$$[Fe^{(n-1)+}- OH \ .... \ \cdot R] \rightarrow ROH \ + \ Fe^{(n-2)+} \tag{45}$$
$$\text{cage}$$

or rebound to a heteroatom (non-oxygen atom) affording R–X,

$$[Fe^{(n-1)+}- OH \ .... \ \cdot R] \xrightarrow{X} RX \ + \ Fe^{(n-1)+}- OH \tag{46}$$

Other pathways include the electron transfer of the incipient radical to yield a carbocation R^+, and desaturation to form olefins,

$$[Fe^{(n-1)+} OH \ .... \ \cdot R] \xrightarrow{-e} [Fe^{(n-1)+}- OH + R^+] \rightarrow R' + [Fe^{(n-2)+}-OH] \tag{47}$$

and radical cage escape,

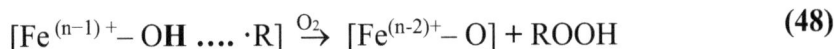

$$[Fe^{(n-1)+}- OH \ .... \ \cdot R] \xrightarrow{O_2} [Fe^{(n-2)+}- O] + ROOH \tag{48}$$

Here, the ratio of the metal-oxo and C-H bond strengths is the major factor that determines the type and direction of the reaction.

The catalytic cycle in the oxidation of the substrate by the participation of ferryl Fe^{n+} ions is presented in Fig. (4) [50].

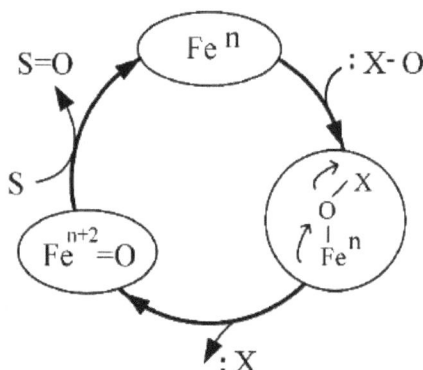

$$S=O \qquad Fe^{n} \qquad :X\text{-}O$$

$$S \qquad \qquad \begin{matrix} X \\ O \\ | \\ Fe^{n} \end{matrix}$$

$$Fe^{n+2}=O \qquad :X$$

Fig. (4). Ferryl-mediated catalytic cycle of an oxidation reaction of the substrate (S), X is a heteroatom (non-oxygen atom), adapted from the ref [50].

Thus, the oxygen atoms of transition metal-oxo organometallic complexes can exhibit two different types of reactions with organic substrates: first, direct oxygen-atom transfer to the substrate; second, hydrogen atom abstraction from the substrate, hydroxylation of the metal ion, and formation of the products [51]. The occurrence of these reactions depends on at least some main factors including the redox potential of the metal center, the strength of the C–H bond broken and the energy of the new bonds formed [46]. The electrophilicity or nucleophilicity of oxo-complexes, which are expressions of the strength of the M-O bonds, may be considered as determining factors in these processes. In the case of the high-valent metal-oxo species, usually, the predominant reaction pathway in the enzymatic activation of the C-H bond is the H-atom abstraction and the oxygen rebound mechanism [52]. The primary, secondary, or tertiary C-H bonds, as well as C-H bonds in saturated and unsaturated hydrocarbons have different electronic structures, therefore, the mechanism of their functionalization can also be different [53].

There is known a correlation between the strength of the metal-oxo bond and the redox potentials of oxo-complexes defined relative to a normal hydrogen electrode (NHE) [54]. Other parameters that are essential for the cleavage of C-H bonds are the equilibrium constants between the metal-oxo and metal-hydroxo (conjugated acid) species, which also correlate with the C-H bond energy data of the substrate. Combining these data in a thermodynamic cycle, the bond energy E_{OH} can be estimated with an error of about 10% (Fig. **5**) [54]

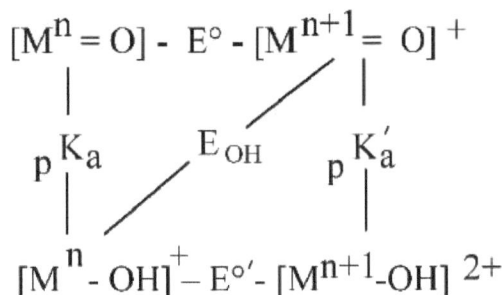

$$[M^n = O] - E° - [M^{n+1} = O]^+$$

$$pK_a \qquad E_{OH} \qquad pK_a'$$

$$[M^n - OH]^+ - E°' - [M^{n+1}-OH]^{2+}$$

Fig. (5). Relationships between the thermodynamic parameters of C-H cleavage of a substrate during its interaction with the metal-oxo moieties [54].

where K_a is the constant of acid-base equilibrium for metal-hydroxo species (conjugate acid for metal-oxo) under the given thermodynamic conditions and pK_a = -lg K_a. E_{OH} is the bond dissociation energy of the OH bonds.

The thermodynamic analysis based on this cycle leads to the following equation:

$$E_{OH} = 23.06 \, E° + pK_a + C$$

where $E°$ is the redox potential of the one-electron reduction of the metal center, C is a constant depending on the nature of the solvent and the redox potential of the system and pK_a = -lgK_a About 10% of the possible inaccuracy may be expected in the estimation of E_{OH} according to this equation in the oxidation of substrates. Applying this equation, Borovik [54] examined manganese-oxo complexes ($[Mn^{IV}H_3buea(O)]^-$ and $[Mn^{IV}H_3buea(O)]^{2-}$ complexes with 9,10-di-hydro-anthracene), and concluded that basicity or acidity plays a crucial role in the C-H bond functionalization in oxidation processes. Data obtained for systems with different metal-oxo species of transition metalorganic complexes show that the energy required to rupture the C-H bond must be comparable to the energy produced to form the MO–H bond [54].

Mechanisms of Catalytic Water Oxidation by the Participation of Transition Metal-oxo Moieties

Water oxidation catalysis by transition metal complexes usually begins with the formation of oxo or oxyl species by different reaction pathways. The summary

reaction is the following

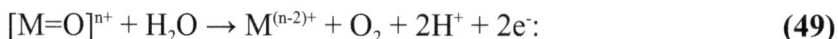

$$[M=O]^{n+} + H_2O \rightarrow M^{(n-2)+} + O_2 + 2H^+ + 2e^-: \tag{49}$$

In this reaction, two different mechanisms have been suggested [55] depending on the predominant resonance form in the metaloxo-metaloxyle equilibrium:

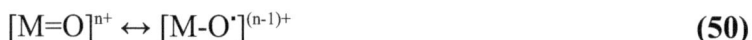

$$[M=O]^{n+} \leftrightarrow [M-O^\bullet]^{(n-1)+} \tag{50}$$

I. According to the *first mechanism* (Scheme **I**), the formation of O-O bond occurs as a result of a nucleophilic attack of either water or hydroxyl group on the oxo moieties [55]. The intermediate is a hydroperoxide, the reactions of intra- or intermolecular deprotonation of which lead to the formation of dioxygen and a reduced metal ion. The final stage is the recovery of the catalyst in a reaction medium, performed by some ways. A brief mechanism of water oxidation, in this case, may be presented by the following scheme consisting of three stages (Scheme **I**, reaction stages 1-3) [55].

Scheme I

1. **a.** Nucleophilic attack of water on the metal-oxo moieties

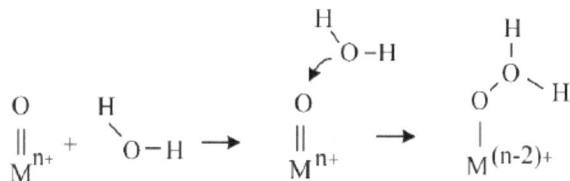

b. Nucleophilic attack of hydroxyl ions on the metal-oxo moieties

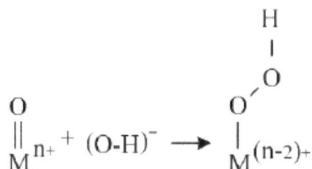

2 **a.** Intermolecular deprotonation assisted by compound A (base)

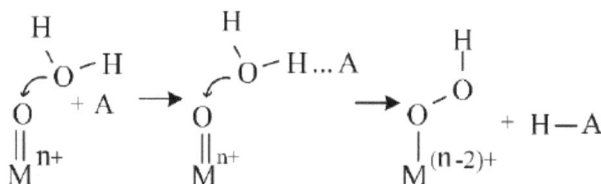

b. Intramolecular deprotonation

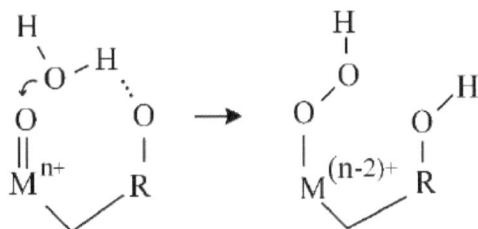

c. Deprotonation of second hydrogen and release of dioxygen

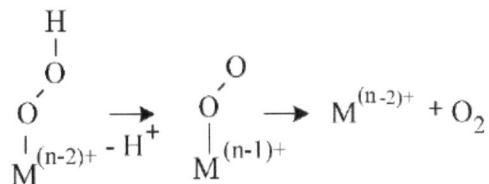

3. Regeneration of the oxo/oxyl moieties of a transition metal

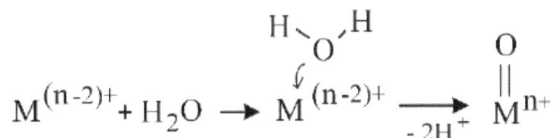

II. According to the *second mechanism* of the water oxidation reaction (Scheme **II**), the radical coupling of two adjacent metal-oxo/oxyl moieties produces peroxidic compound that hydolyses into M-OH and M-OOH in the reaction medium. Consequently, dioxygen is released by the formation of M-O-M (Scheme **II**).

Scheme II

1.

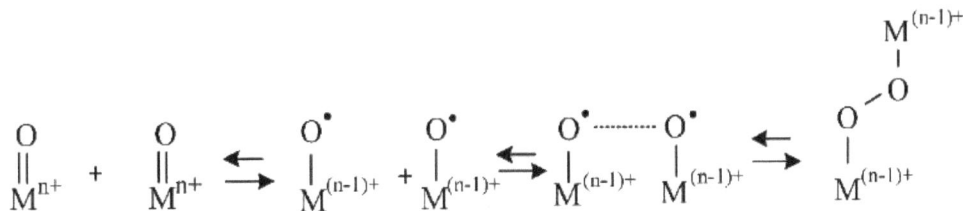

2.

3.

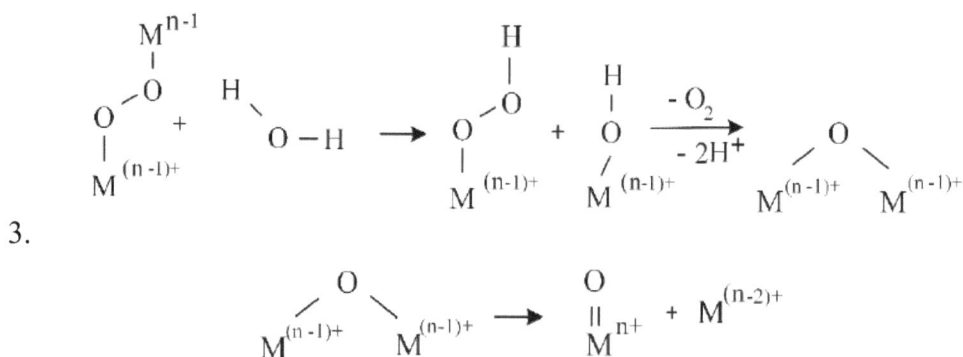

Spin State and Reactivity of Metal-oxo Complexes. Two Spin State and Multiple Spin State Reactivities in the Mechanisms of Oxo-atom Transfer Reactions

The chemical reactivity of oxo-complexes depends on a large number of factors and parameters of the chemical systems and, the first, on the nature of their electronic structure. In this regard, the spin states of chemical entities are fundamental to understanding their reactivity under certain conditions. Historically, the first important attempt to reveal the role of the spin state in the reactivity of reactants was the application of the conservation rules related to the orbital angular momentum and spin in the chemical reactions of diatomic molecules by Wigner and Witmer in 1928 [56]. These rules, based on symmetry considerations, state "that transitions between the terms of the same multiplicity are spin-allowed, while transitions between the terms of different spin-multiplicity are spin-forbidden" [3]. In other words, the chemical reaction may be either spin-allowed or spin-forbidden depending on the molecular terms of the reactants and products. Later Woodward and Hoffmann, as well as Fukui formulated more general rules of the conservation of orbital symmetry in the chemical reactions [57]. These rules make it possible to theoretically predict the occurrence of certain reactions based on the electronic configurations of chemical entities participating in the reaction. In a spin-forbidden reaction, the spin angular momentum is not conserved, and the reactants have a spin multiplicity that differs from the spin multiplicity of the products, while in a spin-allowed reaction, the total spin of the reactants is conserved. However, despite the restrictions imposed by the spin conservation rules, many reactions, formally referred to as spin-forbidden, are ubiquitous [8, 58 - 60]. How can these restrictions be overcome during the reaction?

The same chemical compound or substrate may have different spin states. Each of them has its own potential energy surface for a certain chemical transformation. If the potential energy surface corresponding to the given spin state of the reactants has no points of intersection (crossing) with other potential energy surfaces corresponding to other spin states of the same reactants, the pathway is termed as single-state reactivity (Fig. **6a**) [58, 59].

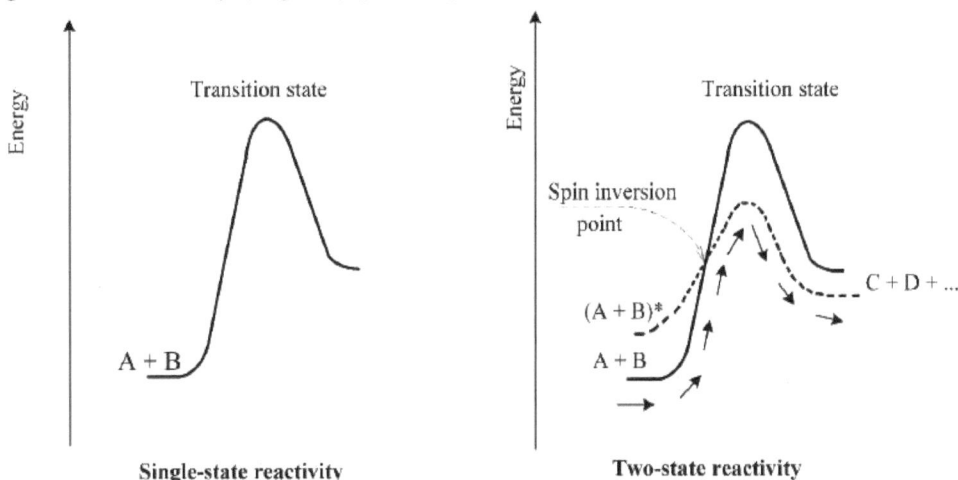

Fig. (6). Reaction profiles showing changes of the potential energy of reactants A+B to products C+D in an endothermic reaction: (**a**) single state reactivity (solid line) ; (**b**) two-state reactivity (solid and dashed lines). (A+B)* is excited state of A+B reactants. The crossing (intersection) point of two profiles is the spin inversion point. Flashes indicate the movement of reactants corresponding to the minimum energy reaction pathway, involving a part of A+B reaction profile (solid line), till the spin inversion point, and a part of (A+B)* reaction profile (dashed line), after this point. Adapted and modified from Ref [59 (p.106)].

If there is at least one point of intersection (crossing) between the two potential energy surfaces, the resulting reaction pathway is known as a two-state reactivity pathway. According to the thermodynamic principle of minimum energy, for a given chemical transformation, the reaction occurs along the minimum energy pathway. Therefore, the initial reaction pathway may change during the reaction, passing to other pathway with a lower potential energy surface, when the potential energy surfaces intersect. The profiles of the two reaction pathways intersecting at one point are shown in Fig. (**6b**). The number of crossing points of two reaction pathways can exceed one. If one or more transitions from one reaction pathway to others take place in the reaction system, the phenomenon is characterized as two- or multi-state reactivity, and the intersection (crossing) point(s) is termed the point(s) of crossing of the minimal energy [58]. As it has been appeared by the DFT calculations [60], in chemical systems, the spin states can be changed during the reaction as nonadiabatic spin inversions, transition from one spin state to another ("rearrangement" of states). This can happen if there is a certain mechanism of interaction for coupling two potential energy surfaces at minimal energy crossing points. The theory of electronic structure does not predict any mechanism of interaction between different potential energy surfaces. In a spin-forbidden reaction, the spin-orbit coupling, as well as spin-spin interactions, can play such a role serving as a "mechanism for coupling of two potential energy surfaces" [58, 60].

The parameters of the induced spin-orbit coupling depend not only on the crystal and ligand fields, but also on the nuclear charge of metal atoms. Therefore, they are more significant and larger for the elements of the third row of the periodic table, than for those of the first and second rows. Usually, the splitting magnitude, originating from the spin-orbit coupling, becomes significant, when the maximum of the vibrational energy of the bond(s) at a given molecular geometry becomes close to the minimum energy at the crossing point of the two potential energy surfaces [60]. Here, the crossing probability, in other words the probability of transition from one spin state to another, can be obtained by quantum mechanical calculations. By inserting the probability parameters into the rate equations, the rate constants can also be calculated and compared with experimental data [60].

In certain cases, the concept of multistate reactivity permits to understand the unusual kinetic features of oxo-atom transfer reactions [58 - 61]. As a classical example of two-state reactivity may be considered the reactions of FeO^+ in the gas phase with different substrates [62, 63]. Let us consider the electronic configuration of this late transition metal-oxo compound. Fe^+ (or $Fe(I)$) ion has the following electronic configuration $[Ar]3d^6 4s^1$. An electronic diagram of the molecular orbitals of FeO^+ is presented in Fig. (7). The molecular ion FeO^+ has 25 electrons and the following electronic configuration of the molecular orbitals [62].

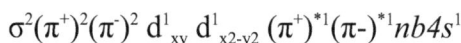

$$\sigma^2(\pi^+)^2(\pi^-)^2 \, d^1_{xy} \, d^1_{x2\text{-}y2} \, (\pi^+)^{*1}(\pi\text{-})^{*1} nb4s^1$$

In [61] it is presented as,

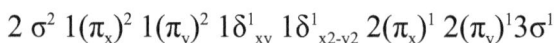

$$2 \, \sigma^2 \, 1(\pi_x)^2 \, 1(\pi_y)^2 \, 1\delta^1_{xy} \, 1\delta^1_{x2\text{-}y2} \, 2(\pi_x)^1 \, 2(\pi_y)^1 3\sigma^1$$

Its ground state molecular term is sextet $^6\Sigma^+$, corresponding to the high-spin state [63].

The electronic configurations of the next higher energy levels relative to the ground state are states with quartet and doublet multiplicity. The lowest energetically excited state (higher than the sextet energy level (ground state), but lower than the doublet state level) has the following electronic configuration:

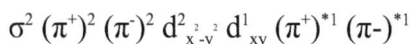

$$\sigma^2 \, (\pi^+)^2 \, (\pi^-)^2 \, d^2_{x^2\text{-}y^2} \, d^1_{xy} \, (\pi^+)^{*1} \, (\pi\text{-})^{*1}$$

This is a quartet spin state (it has the term $^4\Phi$ or $^4\Delta$), and corresponds to the excited low-spin state compared to the ground sextet state. Note that the theoretical predictions [62] and experimental data [63] of the electronic structure of FeO^+ were in satisfactory agreement (Table **2**).

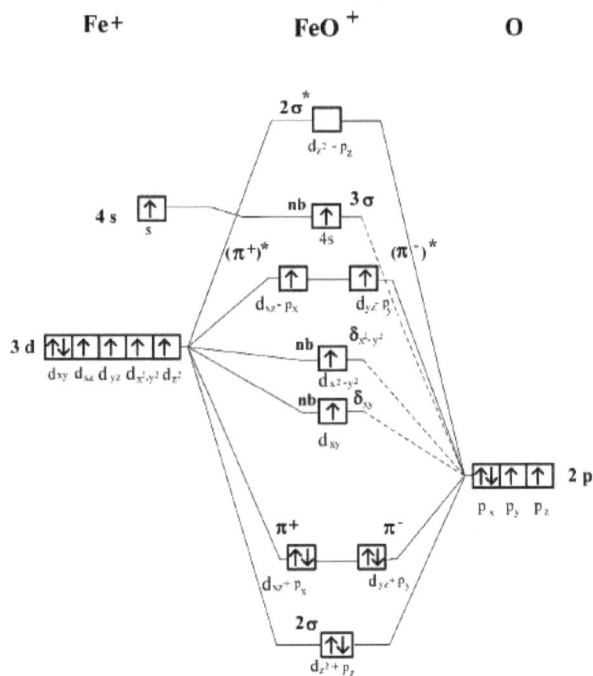

Fig. (7). Energy level diagram of the molecular orbitals of FeO^+ in the ground and high-spin state electronic configuration (nb is a nonbonding molecular orbital). In this diagram the energetic position of 4s-atomic and nb(4s)-molecular orbitals have been chosen arbitrarily [61, p.109].

Table 2. Electronic structure, spin state and molecular terms of FeO^+ ions.

Electronic State	Electronic Configuration	Spin State	Term	Refs.
FeO^+ ground state	$\sigma^2(\pi^+)^2(\pi^-)^2\, d^1_{x^2-y^2}\, d^1_{xy}\, (\pi^+)^{*1}(\pi^-)^{*1} nb4s\sigma^1$	high-spin	$^6\Sigma^+$	[61, 63]
FeO^+ excited state	$\sigma^2(\pi^+)^2(\pi^-)^2\, d^2_{x^2-y^2}\, d^1_{xy}\, (\pi^+)^{*1}\, (\pi^-)^{*1}$	low-spin	$^4\Phi$ or $^4\Delta$	[62, 63]

As it has been mentioned by many authors [59, 61(p.104)], there is an obvious analogy in the reactivity of sextet FeO^+ (high-spin ground state) and triplet dioxygen in reactions with organic compounds. On the other hand, FeO^+ species are radical-ions with the resonance structure $Fe^{+\bullet}-O^\bullet$ rather than molecules with $Fe^+=O$ double bonds. Therefore, in the ground state of FeO^+, the pathway of the stepwise radical reaction in the interaction with organic compounds is significant and often more expected than the concerted reaction with the same substances.

The peculiarities of the reactivity of FeO^+ and other metal-oxo species in oxygen transfer reactions may be obviously exemplified by two reactions: the first with hydrogen [64, 65] and the second with methane [59] or, more generally, with hydrocarbons RH [61(p.104)]. They are relatively more investigated by both theoretical and experimental methods and can be considered prototype reactions in biology from a didactic point of view.

A priori the reactions of FeO^+ both of in the sextet and quartet spin states with the hydrogen molecule in the gas phase are as follows:

$$FeO^+ \, (^6\Sigma^+) + H_2 \rightarrow Fe^+(^6D) + H_2O \tag{49}$$

$$FeO^+ \, (^4\Phi) + H_2 \rightarrow Fe^+(^4F) + H_2O \tag{50}$$

Both reactions 49 and 50 are highly exothermic and, at the first glance, spin-allowed for the transformations of reactants to products. Therefore, it seems that there are no restrictions limiting the rate of these reactions. However, the reaction of FeO^+ with H_2 is very slow. The reaction pathways obtained by quantum mechanical calculations for reactions (49 and 50) are shown in Fig. (**8**).

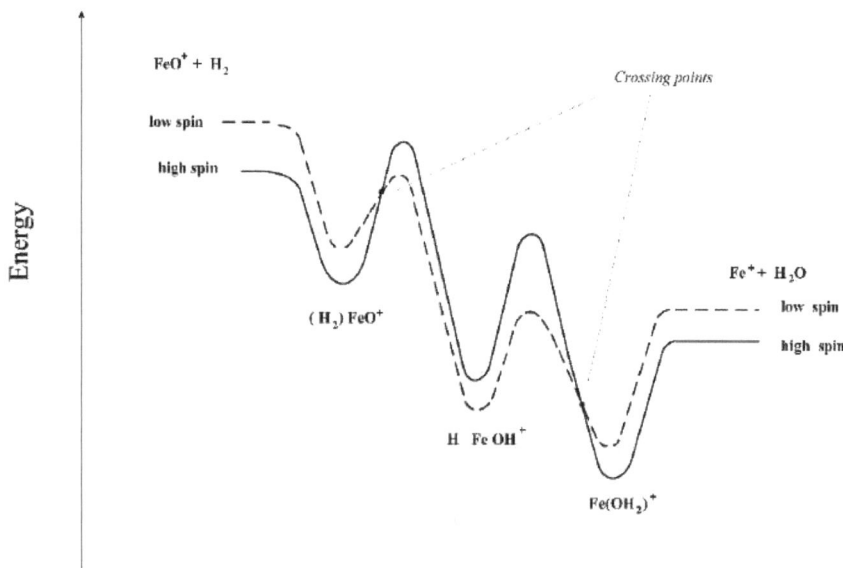

Fig. (8). Energy profiles of the reaction pathways for $FeO^+ + H_2$ system [64, 65].

These two reactions proceed by different mechanisms. The reaction pathway (50), which corresponds to a low-spin state but a relatively high-energy (electronically excited) state of FeO$^+$ ($^4\Phi$), can be started as a concerted and spin-allowed reaction. Conversely, in the initial interaction of FeO$^+$ ($^6\Sigma^+$) and H$_2$ (reaction 49), as it was mentioned above, the reaction can have a radical mechanism. The occurrence of the reaction is related to two changes in the spin states of the reaction system. Based on the presented pathways in Fig. (**8**)., and taking into consideration the thermodynamic principle of minimal energy, the reaction starts as an interaction of high-spin state FeO$^+$ with hydrogen. Then, at the crossing point with the low-spin state pathway, the reaction continues by the pathway corresponding to the system with the low-spin state, changing the mechanism of the interactions. At the second crossing point, the system again changes its spin state and returns to the high-spin state pathway. Therefore, the final product was obtained in the sextet spin state. Thus, according to this description, the overall reaction pathway exhibits two-state reactivity. It is also evident that both changes in the spin states significantly lower the activation barriers during the reaction. Apparently, due to the overcoming of the very high energetic barriers for the reaction (49), the overall reaction of FeO$^+$ with H$_2$ becomes possible, although the extent of the reaction is not high [61]. However, experimental investigations of the gas phase reaction showed that the predominant product of this reaction was the quartet spin state Fe$^+$(^{4}F), but not the sextet spin state Fe$^+$(^{6}D) [65]. It is also obvious that the potential energy surfaces obtained in early quantum mechanical calculations do not completely agree with the experimental kinetic data for this reaction [65].

The formation of the product Fe$^+$ (^{4}F) occurs by the following reaction:

$$FeO^+ \ (^6\Sigma^+) + H_2 \rightarrow Fe^+(^4F) + H_2O \qquad \textbf{(51)}$$

At the same time, it was revealed that in a reaction with high energy reactants, the predominant reaction was the hydrogen atom abstraction with the formation of FeOH$^+$, and at intermediate energies, the radical ("rebound") mechanism was more significant. According to the authors [64], the two-state reactivity in this system is pronounced as a spin inversion in the initial reaction complex from the sextet to quartet and as "a much less efficient quartet-sextet back-inversion in the final reaction complex". In any case, the kinetic data of the reaction (51) indicate the usefulness of the multistate reactivity concept for understanding the peculiarities of the reactivity of a metal-oxo compound in an oxygen atom transfer reaction.

Let us return once again to the above-mentioned analogy in the reactivity of dioxygen and metal-oxo species, from the point of view of the reactions of these reactants with hydrogen. Triplet dioxygen (biradical) does not react with hydrogen in the absence of external factors generating radicals in the system (heat, irradiation, the addition of catalyst or excited species, *etc.*). An isolated gaseous mixture of hydrogen and oxygen can be stored practically infinite under certain conditions. This spin-forbidden reaction occurs as a stepwise process *via* the radical and chain mechanism, if it is initiated in any way. On the other hand, the reaction between hydrogen and singlet oxygen ($^1\Delta$) is a spin-allowed reaction, as both the initial reactants and the product of reaction H_2O are in the singlet state. However, according to the experimental results, in the gas phase, the reaction of O_2 ($^1\Delta$) with hydrogen also occurs predominantly by the chain-radical mechanism, but not as a concerted reaction [66]. Apparently, singlet oxygen participates in the initiation stage of the chain radical process, reducing the induction period of the chain initiation reaction. Thus, in the reaction of singlet O_2 with hydrogen, the possibility of the two-state reactivity and the existence of at least two minimum energy crossing points on the potential energy surfaces may be hypothesized: the first is the transfer from the low-spin state (exited state) pathway to the high-spin state (ground state) pathway; the second is the reverse transfer since the final product is in a singlet spin state. Unfortunately, to the best of our knowledge, there are no extended DFT quantum-mechanical calculations or experimental investigations related to this hypothesis.

These observations are also evidence of the specificity of the electronic structure and reactivity of metal-oxo species. Two-state reactivity was also revealed in the gas phase reaction of FeO^+ cation with methane [59 (p.154)]. The profiles of the potential energy surfaces of sextet and quartet FeO^+ are shown in Fig. (**9**).

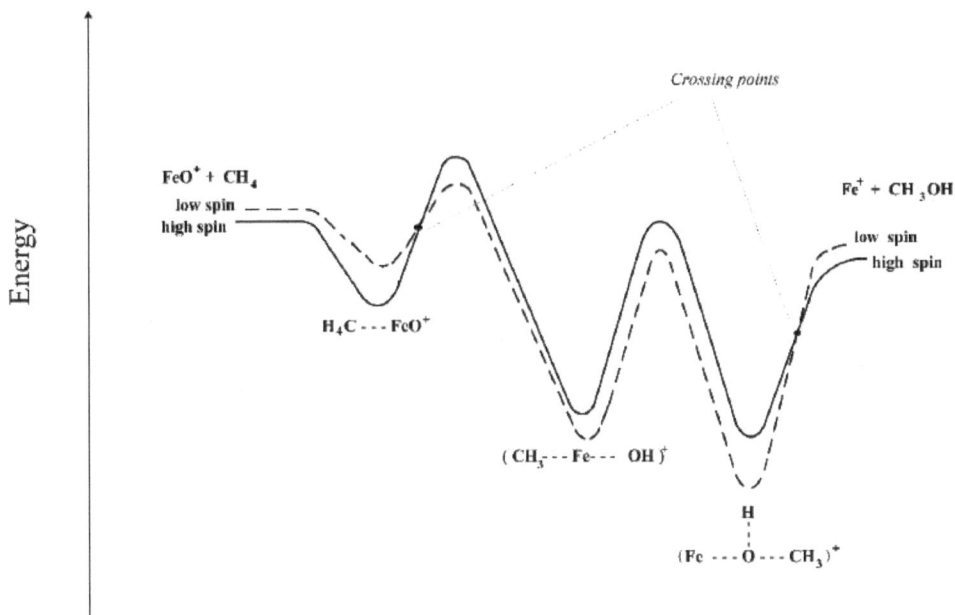

Fig. (9). Energy profiles of reaction pathways for FeO⁺+ CH₄ system [67].

The reaction mechanism for the system $FeO^+ + CH_4$ can be based on two different pathways. Taking into consideration all the above explanations related to the electronic structure of FeO^+, these two pathways for the primary interaction may be represented by the following scheme [61].

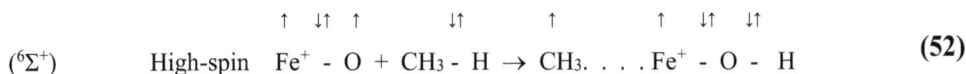

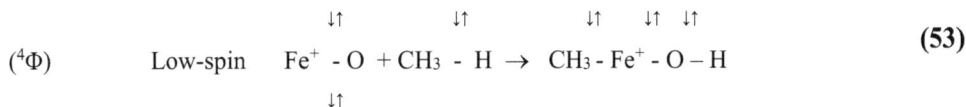

$$
(^6\Sigma^+) \quad \text{High-spin} \quad Fe^+ \text{ - } O + CH_3 \text{ - } H \rightarrow CH_3. \ldots Fe^+ \text{ - } O \text{ - } H \tag{52}
$$

$$
(^4\Phi) \quad \text{Low-spin} \quad Fe^+ \text{ - } O + CH_3 \text{ - } H \rightarrow CH_3 \text{ - } Fe^+ \text{ - } O - H \tag{53}
$$

There are two crossing points on the pathways of reactions 52 and 53, which indicate the changes in the spin states and reaction mechanisms in this system. The computational data show that due to the two-state reactivity phenomenon, the energy barriers of transition states of $(CH_3FeOH)^+$ are significantly lower for reaction 53 than for reaction 52 [67]. The formation of $(CH_3FeOH)^+$ is more probable by a low-spin reaction pathway than by a high-spin pathway, as at a low-spin state, there are energetically low-lying empty molecular orbitals of FeO^+, which can be used to form new chemical bonds. In the sextet state of FeO^+, these molecular orbitals are filled.

In the general case, the phenomenon of two-state or multistate reactivity is exhibited in many reactions of transition metal-oxo compounds with saturated and unsaturated hydrocarbons. For instance, the reaction pathways corresponding to the early transition metal-oxo compounds, such as ScO^+, TiO^+, VO^+, CrO^+, and MnO^+ in reactions with alkanes usually have only one crossing point, corresponding to the spin inversion from a low-spin (ground state) to a high-spin state pathway in the reacting system [59].

SOME CONCLUDING REMARKS ON MECHANISMS OF OXYGEN ATOM TRANSFER REACTIONS

This brief presentation of the mechanisms of oxygen transfer reactions of transition metal-oxo complexes demonstrates their diversity and specificity in the oxidative catalysis of substrates and in the oxidation of water, which are of great theoretical and practical importance in chemistry and biology.

As has been shown above, there are two general types of reaction mechanisms in oxygen atom transfer processes. One of them is based on a direct and, apparently, concerted oxygen atom transfer reaction from the metal-oxo moieties to the substrate, and the second is revealed as stepwise, more often radical, or according to the terminology accepted in this area, "rebound" mechanisms of interaction with the substrates. A more detailed molecular mechanism each of them also has its peculiarities. For instance, the transfer of an oxo-atom can be either an inner sphere or an outer sphere reaction.

A successful catalytic cycle in oxidation processes by the participation of the oxo-atom of transition metals depends not only on the oxidizing ability of the metal-oxo species towards the substrate but also on the re-oxidability of the metal center with the oxidant agents in the reaction medium.

One of the key features of the mechanisms in the transfer reactions of oxo-atoms is the high probability of the appearance of two-state or multi-state reactivity related to the changes in the spine states in a reaction system. This phenomenon has been observed in a great number of oxo-transfer reactions in chemical and, chiefly, biological systems with the participation of transition metal ions or their organometallic complexes in enzymes. It may be considered that one or more spin-crossovers are nearly common for systems involving transition metal oxo-atom transfer reactions.

The main factors determining the rate of oxo-atom transfer reactions and their mechanisms are the driving forces of the processes and intrinsic energy barriers. Therefore, the reaction conditions, on the one hand, and the geometric structure and nature of the ligand, on the other hand, become "tools" that influence or even control the oxidation process.

CONSENT FOR PUBLICATION

Not Applicable.

CONFLICT OF INTEREST

The authors declare no conflict of interest, financial or otherwise.

ACKNOWLEDGEMENT

Declared none.

REFERENCES

[1]   Gray, H.B. Elements of life at the Oxo wall. *Chem. Int.,* **2019**, *41*(4), 16-19.
        [http://dx.doi.org/10.1515/ci-2019-0407]

[2]   Gray, H.B.; Winkler, J.R. Living with Oxygen. *Acc. Chem. Res.,* **2018**, *51*(8), 1850-1857.
        [http://dx.doi.org/10.1021/acs.accounts.8b00245] [PMID: 30016077]

[3]   IUPAC. *Compendium of Chemical Terminology,* 2nd; A.D, McNaught; A, Wilkinson, Eds. Blackwell
        Scientific Publications: Oxford, **2019**.
        [http://dx.doi.org/10.1351/goldbook]

[4]   Holm, R.H. Metal centered oxygen atom transfer reactions. *Chem. Rev.,* **1987**, *87*(6), 1401-1449.
        [http://dx.doi.org/10.1021/cr00082a005]

[5]   Chen, Z.; Yin, G. The reactivity of the active metal oxo and hydroxo intermediates and their
        implications in oxidations. *Chem. Soc. Rev.,* **2015**, *44*(5), 1083-1100.
        [http://dx.doi.org/10.1039/C4CS00244J] [PMID: 25566588]

[6]   Arzoumanian, H. Molybdenum-oxo and peroxo complexes in oxygen atom transfer processes with O$_2$
        as the primary oxidant. *Curr. Inorg. Chem.,* **2011**, *1*(2), 140-145.
        [http://dx.doi.org/10.2174/1877944111101020140]

[7]   Arzoumanian, H.; Bakhtchadjian, R. Oxo-atom transfer reactions of transition metal complexes in
        catalytic oxidation with O$_2$ on the light of some recent results in molybdenum-oxo chemistry.
        Chemical Journal of Armenia (ISSN 0515- 9628), **2012**, *65*, 168-188.

[8]   Teopold, K.H. Dioxygen activation by organometallics of the early transition metals. In:
        Organometallic Oxidation Catalysis (Series: Top Organomet. Chem.,); Meyer, F.; Limberg, C., Eds.;
        Springer: Berlin, **2007**; Vol. 22, pp. 17-39.
        [http://dx.doi.org/10.1007/3418_038]

[9]   Rybak-Akimova, E.V. Mechanism of oxygen binding and activation at transition metal centers
        (Chapter). In: *Physical Inorganic Chemistry: Reactions, Processes, and Applications*; Bakac, A., Ed.;
        Wiley: New Jersey, **2010**; pp. 109-188.
        [http://dx.doi.org/10.1002/9780470602577.ch4]

[10]   Breslow, R. Oxidation by remote functionalization methods (1.3, Oxidation of Unactivated C-H Bond).
        In: *Comprehensive Organic Syntheses, Strategy and Efficiency of Modern Organic Chemistry,
        Oxidation*; Trost, B. M.; Fleming, I., Eds.; Ley, S. V., Volume Ed.; Elsevier Science, Pegamon Press:
        Oxford, New York, Seoul, Tokyo; **1991**; Vol. 7, pp. 39-51.

[11]   Bawden, D.J. Classification of chemical reactions: Potential, possibilities, and continuing relevance. *J.
        Chem. Inf. Comput. Sci.,* **1991**, *31*(2), 212-216.
        [http://dx.doi.org/10.1021/ci00002a006]

[12] Bakhtchadjian, R. *Bimodal Oxidation: Coupling of Heterogeneous and Homogeneous Reactions*; CRC Press, Francis and Taylor: Boca-Raton, New York, London, **2019**.

[13] Sheldon, R.A.; van Bekkum, H. *Fine Chemicals through Heterogeneous Catalysis*; John Wiley and Sons: Weinheim, **2007**, p. 475.

[14] Sheldon, R.A. Heterogeneous catalytic oxidation and fine chemicals. In: *Studies in Surface Science and Catalysis*; Guisnet, M.; Barrault, J.; Bouchoule, C.; Duprez, D.; Perot, G.; Maurel, R.; Montassier, C., Eds.; Elsevier Science B.V.: Amsterdam, **1991**; Vol. 59, pp. 33-54.

[15] Sheldon, R.A.; Arends, I.; Hanefeld, U. *Green Chemistry and Catalysis*; Wiley, VCH: Weinheim, Germany, **2007**, pp. 137-140.
[http://dx.doi.org/10.1002/9783527611003]

[16] Studer, A.; Curran, D.P. *Catalysis of radical reactions: A radical chemistry perspective,* Angewandte Chimie International Edition; **2016**, *55*(1), pp. 58-102.

[17] Liu, X.; Sang, Y.; Yin, H.; Lin, A.; Guo, Z.; Liu, Z. Progress in the mechanism and kinetics of Fenton reaction. *MOJ Eco Environ. Sci,* **2018**; *3*(1), 10-14.
[http://dx.doi.org/10.15406/mojes.2018.03.00060]

[18] Haber, F.; Willstätter, R. Unpaarigheit und Radikalketten im Reaktion Mechanismus organischer und enzymatischer Vorgänge. *Ber. Dtsch. Chem. Ges. B,* **1931**, *64*(11), 2844-2856.
[http://dx.doi.org/10.1002/cber.19310641118]

[19] Haber, F.; Weiss, J. Über die Katalyse des Hydroperoxydes. *Naturwissenschaften,* **1932**, *51*(51), 948-950.
[http://dx.doi.org/10.1007/BF01504715]

[20] Nalbandyan, A.B.; Vardanyan, I.A. *Current State of the Problems of the Gas Phase Oxidation of Organic Compounds*; Izd-vo Acad. Nauk Arm. SSR: Yerevan, **1986**.

[21] Sheldon, R.; Koshi, J.K. *Metal-Catalyzed Oxidations of Organic Compounds: Mechanistic Principles and Synthetic Methodology Including Biochemical Processes*; Academic Press: New York, **1981**, pp. 121-123.

[22] Emanuel, N.M.; Denisov, E.T.; Maizus, Z.K. *Chain Reactions of the Oxidation of Hydrocarbons in Liquid Phase*; Nauka: Moscow, **1965**.

[23] Parkins, A.V. Recent development in platinum group metals catalysis in petrochemical industry. In: *Chemistry of the Platinum Group Metals: Recent Developments (Studies in Inorganic Chemistry, 11)*; Hartley, F.R., Ed.; Elsevier: Amsterdam, **1991**; pp. 106-120.

[24] Clerici, M.G.; Ricci, M.; Strukul, G. Formation of C-O bond by oxidation. In: *Metal-catalysis in Industrial Organic Processes*; Chiusoli, G.P; Maitlis, PM. RSC, Publishing: Cambridge, **2006**; pp. 23-78.

[25] Muzart, J. Palladium-catalyzed oxidation of primary and secondary alcohols. (A review). *Tetrahedron,* **2003**, *59*(31), 5789-5816.
[http://dx.doi.org/10.1016/S0040-4020(03)00866-4]

[26] Singh, H.S. Oxidation of organic Compounds with Osmium Tetroxide. In: *Organic Synthesis by Oxidation with Metal Compounds*; Mijs, W.J.; de Jonge, C.R.H.I., Eds.; Plenum Press: New York, London, **1986**; p. 633.
[http://dx.doi.org/10.1007/978-1-4613-2109-5_12]

[27] Doornkamp, C.; Ponec, V. The universal character of the Mars and Van Krevelen mechanism. *Molecular Catalysis,* **2000**, *162*, 19-32.
[http://dx.doi.org/10.1016/S1381-1169(00)00319-8]

[28] Shilov, A.E.; Shul'pin, G.B. Activation of C-H bonds by metal complexes. *Chem. Rev.,* **1997**, *97*(8), 2879-2932.
[http://dx.doi.org/10.1021/cr9411886] [PMID: 11851481]

[29]   Ye, S.; Ding, C.; Li, C. Artificial photosynthesis systems for catalytic water oxidation. In: *Advances in Inorganic Chemistry*; Elsevier- Academic Press: Cambridge, **2019**; 74, pp. 6-59.

[30]   Schilling, M.; Luber, S. Insights into artificial mater oxidation - A computational perspective. In: *Advances Inorganic Chemistry*; Elsevier-Academic Press (An Imprint of Elsevier): Cambridge, **2019**; 74, pp. 62-108.

[31]   Jurss, J.W.; Concepcion, J.J.; Butler, J.M.; Omberg, K.M.; Baraldo, L.M.; Thompson, D.G.; Lebeau, E.L.; Hornstein, B.; Schoonover, J.R.; Jude, H.; Thompson, J.D.; Dattelbaum, D.M.; Rocha, R.C.; Templeton, J.L.; Meyer, T.J. Electronic structure of the water oxidation catalyst cis,cis-[(bpy)$_2$(H$_2$O)Ru(III)ORu(III)(OH$_2$)(bpy)$_2$]$^{4+}$, the blue dimer. *Inorg. Chem.,* **2012**, *51*(3), 1345-1358.
[http://dx.doi.org/10.1021/ic201521w] [PMID: 22273403]

[32]   Sankaralingam, M.; Lee, Y-M.; Nam, W.; Fukuzum, S. Amphoteric reactivity of metal–oxygen complexes in oxidation reactions. *Coord. Chem. Rev.,* **2018**, *365*, 41-59.
[http://dx.doi.org/10.1016/j.ccr.2018.03.003]

[33]   Shimoyama, Y.; Kojima, T. Metal−oxyl species and their possible roles in chemical oxidations. *Inorg. Chem.,* **2019**, *58*(15), 9517-9542.
[http://dx.doi.org/10.1021/acs.inorgchem.8b03459] [PMID: 31304743]

[34]   Yamaguchi, K.; Takahara, Y.; Fueno, T. Ab-Initio molecular orbital studies of structure and reactivity of transition metal-oxo compounds. In: *Applied Quantum Chemistry*; Smith, V.H., Jr; Scheafer, I.I.I.H.F.; Morokuma, K., Eds.; D. Reidel Publishing Company: Boston, **1986**; pp. 155-184.
[http://dx.doi.org/10.1007/978-94-009-4746-7_11]

[35]   a) Dahl, P.J. Carl Johan Ballhaused **1926-2010**, 1-16.b) Wincler, L.R.; Gray, H.R. Molecular electronic structure of transition metal complexes. In: *Molecular Electronic Structures of Transition Metal Complexes I (Series: Structure and Bonding*; Mingos, D.M.P.; Day, P.; J.P., P., Eds.; Springer-Verlag: Berlin, **1926-2010**; pp. 17-28.

[36]   Crabtree, R.H. The Organometallic Chemistry of the Transition Metals. *Chapter 11. M-L multiple bonds,* (7th ed.); John Wiley and Sons; Hoboken, **2019**, p. 290.

[37]   Cartert, E.A.; Goddard, W. A early-*versus* late-transition-metal-oxo bonds: The electronlc structure of VO$^+$ and RuO$^+$. *J. Phys. Chem.,* **1988**, *92*, 2109-2115.
[http://dx.doi.org/10.1021/j100319a005]

[38]   Ballhausen, C.J.; Gray, H.B. The electronic structure of the vanadyl ion. *Inorg. Chem.,* **1962**, *1*(1), 111-122.
[http://dx.doi.org/10.1021/ic50001a022]

[39]   Chen, Z.; Yin, G. The reactivity of the active metal oxo and hydroxo intermediates and their implications in oxidations. *Chem. Soc. Rev.,* **2015**, *44*(5), 1083-1100.
[http://dx.doi.org/10.1039/C4CS00244J] [PMID: 25566588]

[40]   de Ruiter, G.; Thompson, N.B.; Takase, M.K.; Agapie, T. Intramolecular C–H and C− F bond oxygenation mediated by a putative terminal oxo species in tetranuclear iron complexes. *J. Am. Chem. Soc.,* **2016**, *138*(5), 1486-1489.
[http://dx.doi.org/10.1021/jacs.5b12214] [PMID: 26760217]

[41]   Sahu, S.; Quesne, M. G.; Davies, C. G.; Dürr, M.; Ivanović-Burmazović, I.; Siegler, M. A.; Jameson, G. N. L. Sahu, S.; Quesne, M.G.; Davies, C.G.; Dürr, M.; Ivanović-Burmazović, I.; Siegler, M.A.; Jameson, G.N.; de Visser, S.P.; Goldberg, D.P. Direct observation of a nonheme iron(IV)-oxo complex that mediates aromatic C-F hydroxylation. *J. Am. Chem. Soc.,* **2014**, *136*(39), 13542-13545.
[http://dx.doi.org/10.1021/ja507346t] [PMID: 25246108]

[42]   Keith Woo, L.; Goll, J.G.; Berreau, L.M.; Weaving, R. Oxygen atom transfer reactions of chromium porphyrins: an electronic rationale for oxo transfer *versus* mu-oxo product formation. *J. Am. Chem. Soc.,* **1992**, *114*(19), 7411-7415.

[http://dx.doi.org/10.1021/ja00045a012]

[43]   Abu-Omar, M.M. Oxygen atom transfer. In: *Physical Inorganic Chemistry: Reactions, Processes, and Applications*; Bakac, A., Ed.; Wiley: New Jersey, **2010**; pp. 75-107.
[http://dx.doi.org/10.1002/9780470602577.ch3]

[44]   McKlimont, K.S. Transition Metal Oxo Chemistry, Baran Group Meeting. **2019**.http://baranlab.org

[45]   Bakhtchadjyan, R.; Manucharova, L.A.; Tavadyan, L.A. Organometallic Mo(VI)-complex grafted on TiO_2 as photocatalyst in oxidation of chlorophenyl substituted alkanes with dioxygen. In: *Advances in Chemistry Research*; Vol. 57, Taylor, J. C., Ed.; Nova Science Publishers: New York, **2020**; 57, pp. 163-190.

[46]   Pierre, J-L.; Thomas, F. Homolytic C–H bond cleavage (H-atom transfer): chemistry for a paramount biological process. *C. R. Chim.*, **2005**, *8*(1), 65-74.
[http://dx.doi.org/10.1016/j.crci.2004.09.005]

[47]   Sacramento, J.J.D.; Goldberg, D.P.; Goldberg, D.P. Factors affecting hydrogen atom transfer reactivity of metal–oxo porphyrinoid complexes. *Acc. Chem. Res.*, **2018**, *51*(11), 2641-2652.
[http://dx.doi.org/10.1021/acs.accounts.8b00414] [PMID: 30403479]

[48]   Bím, D.; Maldonado-Domínguez, M.; Rulíšek, L.; Srnec, M. Beyond the classical thermodynamic contributions to hydrogen atom abstraction reactivity. *Proc. Nat. Acad. Sci. USA*, **2018**, *115*(44), E10287-E10294.
[http://dx.doi.org/10.1073/pnas.1806399115] [PMID: 30254163]

[49]   Gao, J. Computation of kinetic isotope effects for enzymatic reactions. *Sci. China Chem.*, **2012**, *54*(12), 1841-1850.
[http://dx.doi.org/10.1007/s11426-011-4433-5] [PMID: 23976893]

[50]   Huang, X.; Groves, J.T. Beyond ferryl-mediated hydroxylation: 40 years of the rebound mechanism and C-H activation. *Eur. J. Biochem.*, **2017**, *22*(2-3), 185-207.
[http://dx.doi.org/10.1007/s00775-016-1414-3] [PMID: 27909920]

[51]   Chen, Z.; Yin, G. The reactivity of the active metal oxo and hydroxo intermediates and their implications in oxidations. *Chem. Soc. Rev.*, **2015**, *44*(5), 1083-1100.
[http://dx.doi.org/10.1039/C4CS00244J] [PMID: 25566588]

[52]   Cho, K-B.; Kang, H.; Woo, J.; Park, Y.J.; Seo, M.S.; Cho, J.; Nam, W. Mechanistic insights into the C-H bond activation of hydrocarbons by chromium(IV) oxo and chromium(III) superoxo complexes. *Inorg. Chem.*, **2014**, *53*(1), 645-652.
[http://dx.doi.org/10.1021/ic402831f] [PMID: 24299279]

[53]   Hartwig, J.F. Evolution of C–H bond functionalization from methane to methodology. *J. Am. Chem. Soc.*, **2016**, *138*(1), 2-24, 2-24.
[http://dx.doi.org/10.1021/jacs.5b08707] [PMID: 26566092]

[54]   Borovik, A.S. Role of metal-oxo complexes in the cleavage of C-H bonds. *Chem. Soc. Rev.*, **2011**, *40*(4), 1870-1874.
[http://dx.doi.org/10.1039/c0cs00165a] [PMID: 21365079]

[55]   Schilling, M.; Luber, S. Computational Modeling of Cobalt-Based Water Oxidation: Current Status and Future Challenges. *Front Chem.*, **2018**, *6*, 100.
[http://dx.doi.org/10.3389/fchem.2018.00100] [PMID: 29721491]

[56]   Hargittai, I. Pioneering quantum chemistry in concert with experiment. In: *Pioneers of Quantum Chemistry*; Strom, E. T.; Wilson, A.K., Eds.; American Chemical Society, Symposium Series. Oxford University Press: Oxford, **2013**; 1122, pp. 47-73.

[57]   Hargittai, M.; Hargittai, I. *Symmetry through the Eyes of a Chemist,* 3rd ed; Springer Science +Business Media B.V, **2009**.
[http://dx.doi.org/10.1007/978-1-4020-5628-4]

[58]   Schröder, D.; Shaik, S.; Schwarz, H. Two-state reactivity as a new concept in organometallic chemistry. *Acc. Chem. Res.,* **2000**, *33*(3), 139-145.
[http://dx.doi.org/10.1021/ar990028j] [PMID: 10727203]

[59a]   Swart, M.; Costas, M. General introduction to spin state (Chapter 1, pp.1-4).

[b]   Dzik, W. I.; Böhmer, W.; de Bruin, B. Multiple spin state scenarios in organometallic reactivity (Chapter 6, pp. 103-126).

[c]   Roithova, J. Multiple spin state Scenarios in gas-phase reactions.(Chapter 8, pp.157-178).

[d]   Quesne, M.; Faponle, A. S.; Goldberg, D. P.; de Visser, S. Catalytic function and mechanism of heme and nonhem-irone(IV)-oxo complexes in nature (Chapter 9, pp. 185-202).

[e]   Cook, S.A.; Lacy, D. C.; Borovik, A. Terminal metal-oxo species with unusual spin states Chapter 10, pp. 203-221); In: Spin States in Biochemistry and Inorganic Chemistry: Influence on Structure and Reactivity: Swart, M.; Costas, M.; Eds.; John Wiley & Sons, Ltd: Chichester, UK; 2015.
[http://dx.doi.org/10.1002/9781118898277]

[60]   Matsunaga, N.; Koseki, S. Modeling of spin-forbidden reactions. In: *Reviews in Computational Chemistry*; KB, Lipkowitz; R, Larter; TR, Cundari, Eds.; Wiley-VCH: Hoboken, NJ, **2004**; 20, pp. 101-145.

[61]   Schroder, D.; Schwaez, H.; Shaik, S. Characterization, orbital description and reactivity patterns transition–metal oxo species in the gas phase. In: *Metal-oxo and Metal-Peroxo Species in Catalytic Oxidations*; Meunier, B., Ed.; Springer: Heidelberg, **2000**; pp. 91-124.
[http://dx.doi.org/10.1007/3-540-46592-8_4]

[62]   Sakellaris, C.N.; Miliordos, E.; Mavridis, A. First principles study of the ground and excited states of FeO, FeO⁺, and FeO⁽⁻⁾. *Chem. Phys.,* **2011**, *134*(23), 234308.
[http://dx.doi.org/10.1063/1.3598529] [PMID: 21702557]

[63]   Halfen, D.T.; Ziurys, L.M. Millimeter/submillimeter velocity modulation spectroscopy of FeO⁺ (X⁶ Σ⁺): Characterizing metal oxide cations. *Chemical Physics Letters,* **2010**, *496*, 8-13.

[64]   Ard, S.G.; Johnson, R.S.; Melko, J.J.; Martinez, O.; Shuman, N.S.; Ushakov, V.G.; Guo, H.; Troe, J.; Viggiano, A.A. Spin-inversion and spin-selection in the reactions FeO⁺ + H₂ and Fe⁺ + N₂O. *Phys. Chem. Chem. Phys.,* **2015**, *17*(30), 19709-19717.
[http://dx.doi.org/10.1039/C5CP01418B] [PMID: 26129708]

[65]   Essafi, S.; Tew, D.P.; Harvey, J.N. The dynamics of the reaction of FeO ⁺ and H₂: A model for inorganic oxidation. *Angew. Chem. Int. Ed. Engl.,* **2017**, *56*(21), 5790-5794.
[http://dx.doi.org/10.1002/anie.201702009] [PMID: 28429418]

[66]   Starik, A.M.; Titova, N.S. Initiation of combustion and detonation in H₂ + O₂ mixtures by excitation of electronic states of oxygen molecules. In: *High Speed Deflagration and Detonation. Fundamentals and Control*; Roy, G.; Frolov, S.; Netzer, D.; Borisov, A., Eds.; ELEX-KM Publishers: Moscow, **2001**; pp. 63-78.

[67]   Jaccob, M.; Sankaralingam, M.; Britto, N.J.J. Activation of small molecules by transition-metal complexes. In: *Chemical Modelling*; Springborg, M.; Joswig, J-O., Eds.; Royal Society of Chemistry: Croydon, UK, **2020**; Vol. 15, pp. 31-172.

CHAPTER 2

Bio-Inspired Dioxygen Activation and Catalysis By Redox Metal Complexes

Guangjian Liao[1] and Guochuan Yin[1,*]

[1] *School of Chemistry and Chemical Engineering, Huazhong University of Science and Technology, Wuhan 430074, PR China*

Abstract: In nature, redox enzymes mediated dioxygen activation with oxidations proceeds smoothly and highly selectively under ambient temperature, whereas in the chemical industry, versatile oxidations are commonly performed at elevated temperature, which leads to the occurrence of radical chain process, thus causing low product selectivity and environmental pollution. This chapter will first introduce the strategies of enzymes including P450s, methane monooxygenase, dioxygenases in dioxygen activation and catalysis, thus illustrating how enzymes activate dioxygen and selectively transfer the resulting active oxygen to their substrates. Then, inspired by enzymatic dioxygen activation, the progress in biomimetic dioxygen activation with related catalytic oxidations by synthetic redox metal complexes will be presented, and its current challenges will be discussed as well. Finally, a recent new strategy for dioxygen activation and catalysis, that is, Lewis acid promoted dioxygen activation by redox metal complexes, will be introduced; this new strategy may have more closely biomimicked enzymatic dioxygen activation than those traditional strategies, thus shedding new light on catalyst design for industrial oxidations.

Keywords: Bio-inspired O_2 activation, Catalytic oxidation, Enzymatic O_2 activation.

INTRODUCTION

Oxidation is one of the most significant processes in nature and the chemical industry. In biological cells, versatile redox enzymes can highly efficiently activate dioxygen and transfer the resulting active oxygen to the substrate with high selectivity at ambient temperature. To achieve this target, the co-enzymes, the active sites of proteins combined together with the electron transfer chain, if needed, synergistically catalyze dioxygen activation and oxygen transfer. However, in the chemical industry, most of the oxidations are performed at eleva-

---

* **Corresponding author Guochuan Yin**: School of Chemistry and Chemical Engineering, Huazhong University of Science and Technology, Wuhan 430074, PR China; E-mail: gyin@hust.edu.cn

ted temperature for achieving reasonable oxidation efficiency, and radical chain processes are hardly avoided, which leads to the low selectivity of the targeted products with resource loss and environmental pollution. Even more seriously, the stoichiometric oxidants are employed in certain cases. For example, the production of adipic acid, a significant monomer for polyamide-6,6 synthesis, represents one of the most important, but also most polluted oxidations in the industry [1]. Commercially, adipic acid is mainly produced through air oxidation of cyclohexane to cyclohexanol and cyclohexanone (KA oil), followed by nitric acid oxidation. The first process in Dupont's is performed at 155-165 °C and 8-10 atm with Co-Mn catalyst, and the oxidation proceeds through a classic radical chain process [2]. To achieve an 85% selectivity of KA oil, the conversion of cyclohexane needs to be controlled below 5-7%. The next nitric acid oxidation of KA oil can achieve 95% selectivity with 100% conversion by using Cu-V catalyst, however, it suffers severe drawbacks including the use of corrosive nitric acid as the oxidant and the unavoidably leaching of N_2O to the atmosphere, which contributed a significant content in the global N_2O emissions. Up to now, industrial oxidations still face serious challenges in dioxygen activation and next selective oxygen transfer to the substrate when compared with those biological oxidations. Currently, the catalytic cycles of dioxygen activation and oxygenation by redox enzymes have been mostly interpreted for some significant oxidation events in nature, and many bio-inspired strategies for dioxygen activation have also been explored for chemical oxidations [3, 4], however, the applications for these biomimetric oxidations in large scale industry is still scarce. This chapter provides a brief summary of the dioxygen activation mechanisms of some popular enzymes with their inspired catalytic oxidations by redox metal complexes.

Dioxygen Activation and Catalytic Oxygenation by Cytochrome P450, Methane Monooxygenases With Their Synthetic Models

1) P450 Mechanism

The cytochrome P450 enzymes represent a superfamily of hemoproteins, which are responsible for the metabolism of xenobiotics and the biosynthesis of critical signaling molecules used for control of development and homeostasis [5, 6]. The active site of the P450 enzymes consists of a heme-iron with a fifth proximal Cys ligand, and the sixth coordination site of the iron center is the site for dioxygen activation with related oxygenation reaction. A simplified mechanism for P450 mediated substrate hydroxylation is illustrated in Scheme **1**. The catalytic cycle starts from the substrate RH binding to the resting state of P450, that is a heme-iron(III) complex, which triggers the change of the spin state of the iron(III) from LS to HS. Next, the electron transfer from NAD(P)H to the iron(III) reduces it to iron(II). Dioxygen activation by the resulting iron(II) generates an iron(III)-OO⁻

superoxo radical species. Further electron and proton transfer to this iron(III)-OO⁻ species yields an iron(III)-OOH⁻ species, which is called the compound 0. Then, the proton assisted hetero-cleavage of the peroxide leads to the formation of the iron(IV)=O⁺˙ cation radical intermediate, which is called the compound I, an formal iron(V)=O species. This iron(IV)=O⁺˙ functions as the key active species for coming substrate oxidations. After the hydrogen atom abstraction from the substrate RH by the iron(IV)=O⁺˙, it generates a substrate radical, R˙, with the iron(IV)-OH species, which is called compound II. Next, the iron(IV)-OH transfers the OH group to R˙ to give the resting state of the catalytic cycle, the heme-iron(III) complex, with the oxygenation product ROH. This oxygenation mechanism by compound I is called as oxygen rebound mechanism, which was coined by Groves [7]. We may see from Scheme **1**, and will further discuss it in the future, that the electron transfer from NAD(P)H plays a key role in triggering dioxygen activation through reducing the iron(III) to the corresponding iron(II). The hydrogen bond network around the active site also play a crucial role in stabilizing the resulting iron(III)-OO⁻ superoxo radical species after dioxygen activation, and in driving the next O-O bond cleavage in the iron(III)-OOH species to generate the compound I, which is responsible for substrate oxidation. In addition, you may also see that P450 enzymes are monooxygenases, which transfer one oxygen atom from dioxygen to their substrates to generate the oxygenation product with another oxygen atom from dioxygen released into the surrounding water.

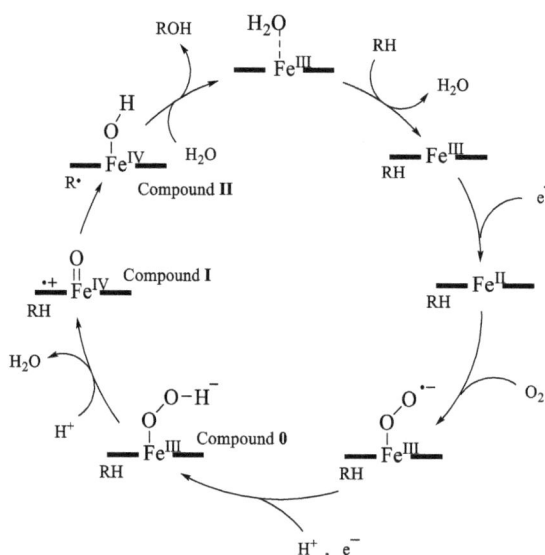

Scheme (1). A simplified catalytic cycle for P450 mediated substrate hydroxylation reaction.

2) sMMO and pMMO Mechanism

The methane monooxygenases are another category of monooxygenases, which are responsible for methane hydroxylation to methanol in nature [8]. There exist two categories of methane monooxygenases including soluble and particulate methane monooxygenases, abbreviated as sMMO and pMMO. Unlike the P450s, in the active site of sMMO, it consists of a diiron cluster having carboxylate bridge from glumate residue in the protein. A simplified catalytic cycle of sMMO is illustrated in Scheme 2 [9], and the catalysis also starts from the electron transfer from NADH to the oxidized diiron(III) state for reducing it to the diiron(II) state, which further activates dioxygen to give a putative $Fe^{II}Fe^{III}$ superoxo species, and next evolves to a peroxo bridged diiron(III) cluster, called the intermediate P*. Then, proton transfer followed by PCET transforms the intermediate P* to the intermediate Q, which was popularly identified by EXAFS as a diamond diiron(IV) core having two μ-oxo bridges. This intermediate Q plays the key role in methane hydroxylation as well as the compound I in P450. Although the identified structure by EXAFS for the intermediates Q was in a diamond diiron(IV) core, the active structure of the intermediate Q for methane hydroxylation is still in question, and the latest evidence implicated that it could be in an open-shell rather than a close-shell structure [10]. Further discussion about the reactivity of the close-shell and open-shell structure in oxidation will be presented later.

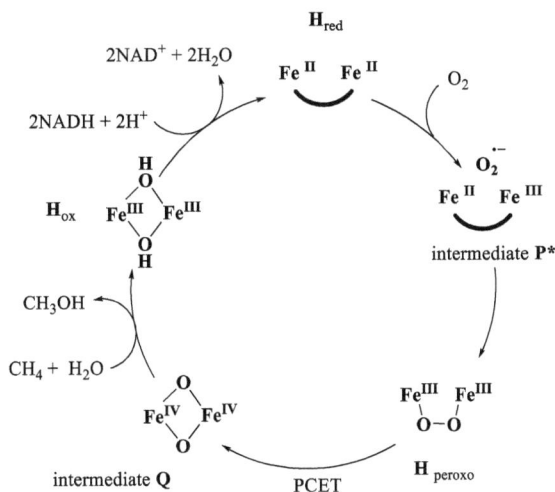

Scheme (2). A simplfied mechanism for sMMO mediated methane hydroxylation.

While the mechanism of sMMO has been relatively well-understood, the mechanism of methane hydroxylation by pMMO is far less understood yet [11]. At the active center of pMMO, there exists one single copper site and another dicopper site. Although the clear evidence is still not enough, it generally prefers that methane hydroxylation happens in the dicopper site. Up to now, the active intermediate for methane hydroxylation is still in argument, in which both the mixed-valent oxodicopper(II/III) and simple oxodicopper(II) centers were proposed as the candidates [12]. It is worth mentioning that pMMO , as well as sMMO, needs an electron transfer chain to provide external electrons for reducing the plausible dicopper(II) cluster, if it was the active center for methane hydroxylation, to a dicopper(I) cluster prior to dioxygen activation.

3) Coreductant Involved Catalytic Oxidation

Because of the high efficiency of monooxygenases with their powerful oxidizing capability, their biomimetics abstracted much attention with the proceeding of their mechanism studies. As shown in the above mechanisms of monooxygenases, to achieve efficient catalysis, three key items are essential, including the active center, like heme-iron in P450 and diiron cluster in sMMO, electron supplier, that is, NAD(P)H, and dioxygen. In biomimetics, Tabushi demonstrated an early example of aerobic cyclohexene oxygenation with manganese(III) porphyrin complex as a catalyst in the presence of $NaBH_4$, which selectively provided cyclohexanol as the product [13]. In this system, $NaBH_4$ functioned the role of NAD(P)H in monooxygenases to reduce their Mn(III) catalyst to the corresponding reduced Mn(II) complex, which triggers dioxygen activation as that in P450. The formation of cyclohexanol as the product was rationalized by that P450-type oxygenation of cyclohexene by their catalysis yielded cyclohexene oxide as the product which was further reduced to cyclohexanol in the presence of $NaBH_4$. Mansuy demonstrated another example of P450-type oxygenation with ascorbate as the electron supplier, which achieved efficient alkane hydroxylation and olefin epoxidation in benzene at 20°C [14]. The authors found that ascorbate not only played the role of NAD(P)H in P450, but also acted as an efficient inhibitor of the autoxidation, since in the absence of ascorbate, oxidation of cyclohexene provided cyclohexenone, cyclohexenol and epoxycyclohexane in a ratio of 79:20:1, a characteristic of autoxidation, while, with ascorbate, it provided solely epoxycyclohexane as the product. Using acetaldehyde as an additive, Murahashi demonstrated a highly efficient oxidation of cyclohexane with only 2.5×10^{-4} mol% of metalloporphyrin catalyst under O_2 atmosphere in ethyl acetate [15]. Unlike the above introduced oxidations, here, the role of acetaldehyde was explained as to generate peracid, which next oxidized the metalloporphyrin catalyst to its active oxometal species for oxygenation, a route of peroxide shunt in the P450 cycle.

Clearly, as biomimetics of monooxygenases, using exogenous electron supplier to trigger the dioxygen activation is the key step for efficient catalysis, because the resting state of the catalyst in a chemical system is generally in a stable form which cannot bind and next activate dioxygen as well as those in monooxygenases. The challenge is that such a strategy of using sacrificial reductant in a catalytic oxidation process is extremely expensive in the industry, except the targeted product was highly valued. Alternatively, to avoid the employment of sacrificial reductant for dioxygen activation in P450 type oxygenation, pre-activated oxidants like hydrogen peroxide are popularly employed as the terminal oxidants for redox metal complexes mediated catalytic oxidations [16], a peroxide shunt in the P450 cycle, which is not be covered in this chapter.

4) The Reactivity of sMMO Like Diamond Core and its Lewis Acid Promoted Oxygenation

The methane monooxygenases represent another category of redox enzymes that demonstrates powerful oxidizing capability, that is, direct hydroxylation of methane to methanol. In the chemical process, such an oxidation is still called the Holy Grail in chemistry. Currently, the industrial production of methanol is performed by first partial oxidation of methane to syngas ($CO+H_2$) followed by catalytic reduction, a process of low atom efficiency with low energy efficiency. Therefore, the identified diiron(IV) diamond core of the intermediate Q for methane hydroxylation is quite attractive, however, the active structure of the intermediate Q is in question up to now. In the DFT calculations carried out by Siegbahn and Crabtree, it was found that the close-shell structure of diiron(IV) core can further evolve to an open-shell structure (Scheme **3**), that is, Fe^{III}-O-Fe^V=O, which can abstract hydrogen atom from methane *via* a low-energy transition state [17]. In the synthetic model, Que demonstrated that the presence of substrate may trigger the collapse of the [$Fe^{III}Fe^{IV}(\mu$-O$)_2$] diamond core to an open-shell Fe^{III}-O-Fe^{IV}=O clusters, and the latter was much more reactive than the close-shell precursor, demonstrating a million-fold faster in C-H bond cleavage [18, 19]. Yin also found that the presence of a Lewis acid like Sc^{3+} can trigger the collapse of *in-situ* generated [$Mn^{IV}_2(\mu$-O$)_2$] diamond core in a catalytic process to a Lewis acid adduct of monomeric O=Mn^{IV} species [20, 21]. The latter was highly active for olefin oxygenation, whereas the diamond precursor was very sluggish. In an iron(II) complex catalyzed olefin epoxidation with H_2O_2, Yin also found that the presence of Lewis acid can substantially improve the catalytic efficiency in olefin oxygenation, and further DFT calculations disclosed an O=Fe^{IV}-O-Sc^{3+} species, generated *in situ*, was the active species for oxygenation. The occurrence of this Lewis acid adduct of the iron(IV)=O species substantially decreased the activation energy barrier of the iron(IV)=O species in olefin epoxidation from

24.8 to 12.2 kcal/mol, whereas in the absence of Sc^{3+}, the oxygenation proceeded by the iron(V)=O species, which has an activation energy barrier of 17.9 kcal/mol [22]. Thus, the presence of a Lewis acid like Sc^{3+} shifted the active species for oxygenation from the oxidation state of iron(V) to that of iron(IV), thus also substantially decreasing the barrier for oxidation of the catalyst to its active state. Notably, the DFT calculations also disclosed that the linkage of Lewis acid like Sc^{3+} to the Fe^{IV}=O unit caused the natural bond orbital (NBO) charge on Fe increasing from +1.05 to +1.14, thus making it more electrophilic, accordingly more active in oxygenation.

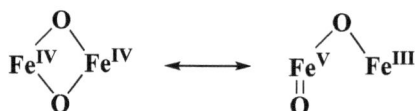

Scheme (3). The simplified close-shell and open-shell structure of the intermediate Q in sMMO.

Particularly, the latest HERFD-EXAFS characterizations of the intermediate Q in sMMO by DeBeer disclosed that the distance of Fe-Fe in Q was 3.4 Å, a distance observed in open-shell synthetic models [10]. In viewing the sharply different reactivity between the close-shell and open-shell unit in synthetic models, an open-shell structure of the intermediate Q in sMMO is highly possible for methane hydroxylation, since the C-H bond in methane is the most robust one in all of those C-H bonds. In addition, the open-shell structure of the intermediate Q is in a Fe^{III}-O-Fe^{V}=O form in the DFT calculations. Apparently, methane hydroxylation occurs on the Fe^{V}=O site, while the Fe^{III} site may modulate the reactivity of the Fe^{V}=O moiety, which resembles the recent works on Lewis acid promoted catalytic oxidations as introduced above. Notably, the studies on Lewis acid modulated oxidative reactivity of active metal ions have abstracted much attention in recent years [23, 24]. The current data disclosed that the interaction of the Lewis acid with the active metal ions may positively shift their redox potentials, thus accelerating their rates in electron transfer. Consisting with the improved electron transfer capability, even in the oxygenation process, the presence of Lewis acid can shift the direct oxygenation of an active oxometal moiety to electron transfer followed by oxygen transfer, and in hydrogen abstraction process, it can also shift a hydrogen atom transfer process of an active oxometal moiety to electron transfer followed by proton transfer. These Lewis acid modulated reactivity changes of the active metal ions have inspired a new strategy for redox catalyst design, which has been applied in versatile oxidation processes [20 - 22, 25 - 27].

Dioxygen Activation by Dioxygenases

1) Dioxygenase Mechanism

Different from the monooxygenases, the dioxygenases transfer both oxygen atoms from dioxygen into the substrate or a primary substrate plus a co-substrate. Even for dioxygen activation, it could also be different from monooxygenases; dioxygenases either take all of the four electrons needed for dioxygen reduction from the substrate, or two from the substrate with two from external electron donor, for example, NADH. Tryptophan 2,3-dioxygenase (TDO) and indoleamine 2,3 dioxygenase (IDO) are the heme-iron dioxygenases that cleave the pyrrole ring of L-tryptophan and insert both oxygen atoms from dioxygen into the substrate. Several mechanisms were even proposed for TDO and IDO oxygenation, and the one based on DFT calculations by Morokuma received the most attention, which is shown in Scheme **4** [28]. In this mechanism, the heme-iron(II) in the active site first activates dioxygen to generate an iron(III) superoxo radical, which attacks the 2-position carbon in pyrrole ring to generate a C=C bond broken substrate radical intermediate. Next, the cleavage of the O-O bond in iron(III)-O-OR gives an epoxide intermediate with the formation of a heme-iron(IV)=O species. Then, an acid-assisted ring-opening of the epoxide yields the next intermediate with the carbon cation at the C2 position which is next attacked by the nearby iron(IV)=O species. The C-C cleavage between C2-C3 position of this intermediate gives the dioxygenation product with the release of the heme-iron(II) center to achieve the catalytic cycle. As shown, both the two oxygen atoms from dioxygen are incorporated into the product, and no external electron is needed for dioxygen activation by TDO dioxygenases.

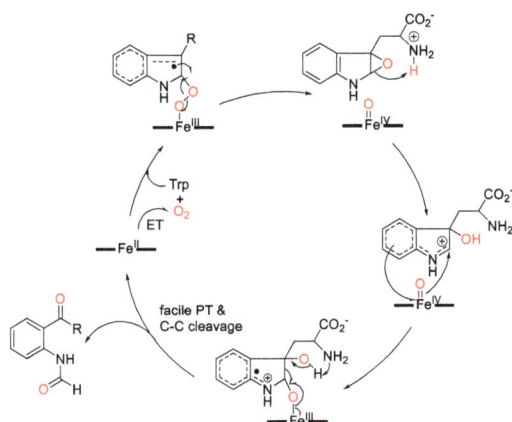

Scheme (4). Proposed dioxygenation mechanism of TDO and IDO based on DFT calculations.

Unlike TDO and IDO dioxygenases, catechol dioxygenases are nonheme-iron containing enzymes, and they have two distinct dioxygenation activities [29]. The extradiol dioxygenases start the catalytic cycle with an iron(II) center, whereas the intradiol dioxygenases start with an iron(III) center (Schemes **5** and **6**). In extradiol dioxygenases, the iron(II) ion is coordinated with two histidines, one glutamate, and one hydroxide with two water ligands in an octahedral structure. The catalysis starts from catechol coordinating to the iron(II) center by the bidentate mode, followed by dioxygen activation to generate an iron(III) superoxo radical species. The intra-molecular electron transfer of this catechol ligated iron(III) superoxo radical intermediate yields an iron(II) superoxo radical species with another radical located at the aromatic ring. Then recombination of two radicals (carbon radical at the aromatic ring and the superoxo radical at the oxygen) in this intermediate yields an alkylperoxo intermediate. Next, the base assisted decomposition of this alkyperoxo intermediate with the O-O cleavage of the alkylperoxide gives a 2,3-epoxide intermediate with the original two oxygen atoms from catechol still ligated to the iron(III) center. The epoxide undergoes a rearrangement to form a seven-membered ε-lactone intermediate which is still ligated to the reduced iron(II)-OH moiety. Finally, catalyzed by the nearby iron(II)-OH moiety, hydrolysis of this ε-lactone intermediate produces the extradiol product with the release of the iron(II) species to achieve the catalytic cycle, meanwhile, the hydroxide from the iron(II), in which its oxygen atom was originally from dioxygen, was incorporated into the extradiol product, achieving the dioxygenation activity.

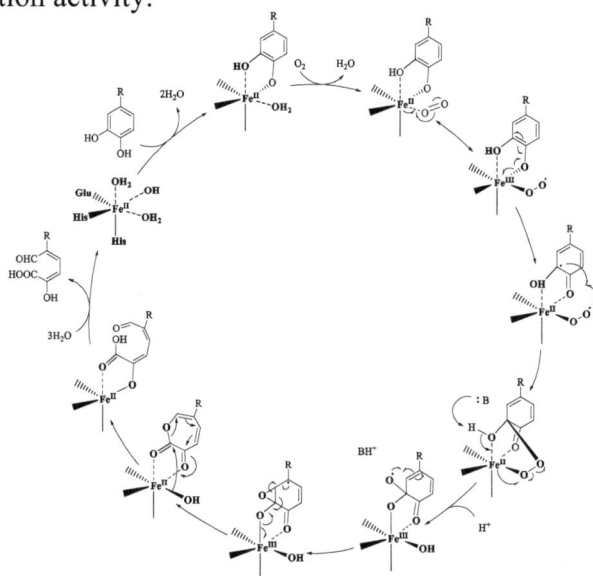

Scheme (5). Proposed catalytic cycle for extradiol ring-cleaving dioxygenases.

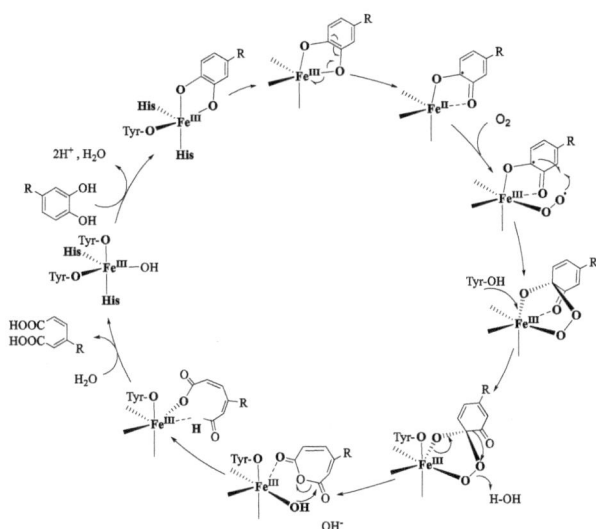

Scheme (6). Proposed catalytic cycle for intradiol ring-cleaving dioxygenases.

Different from the extradiol dioxygenases, the intradiol dioxygenases have an iron(III) center before the start of the catalytic cycle (Scheme **6**) [29]. The iron(III) center is coordinated with two histidine and two tyrosinate residues with the fifth hydroxide in a trigonal bipyramidal structure, which is different from the octahedral structure in the extradiol dioxygenases. Apparently, the iron(III) species is not able to activate dioxygen directly. To initiate the catalysis, catechol binds to the iron(III) center by a bidentate mode with the release of one tyrosinate residue. Next, the intra-molecular electron transfer of this catechol bound iron(III) species yields the iron(II) center with a radical located at the aromatic ring of catechol. Next, dioxygen activation by the resulting iron(II) center generates the iron(III) superoxo radical intermediate, which undergoes intramolecular radical recombination, including carbon radical at the aromatic ring and the superoxo radical at the oxygen, to produce the alkylperoxo intermediate. The collapse of this alkylperoxo intermediate with the O-O bond cleavage occurs *via* a Crigge rearrangement to yield a cyclic anhydride intermediate, meanwhile, an iron(III)-OH moiety was generated. Finally, the iron(III)-OH catalyzed hydrolysis of this anhydride generates the muconic acid product, which is different from the extradiol product, with the release of the iron(III) center to achieve catalytic cycle.

While the alkylperoxo intermediates occur in both extradiol and introdiol dioxygenases, the cleavage of the aromatic ring happens differently at this stage, possibly due to their difference in the detailed structures in the alkylperoxo intermediate. In extradiol dioxygenases, both of the catecholic oxygen atoms bound to the iron(II) center, whereas in intradiol dioxygenases, one of the oxygen atoms from catechol is released from the coordination sphere of the iron(III) center. This structural difference changes the alignment of the O-O bond of the peroxo with the bonds of the aromatic ring, thus altering the insertion site of oxygen after the O-O bond cleavage. The extradiol cleavage occurs *via* 1,2-alkenyl migration to give a lactone, whereas the intradiol cleavage occurs *via* 1,2-acyl migration to give an anhydride, thus leading to different oxygenation products. In addition, the intradiol dioxygenases are special ones, which start the catalytic cycle from an iron(III) center, and they utilize the coming catechol substrate to reduce the iron(III) species *via* intra-molecular electron transfer to the iron(II) moiety for dioxygen activation. A similar strategy for dioxygen activation occurs in flavonol 2,4-dioxygeanses, in which the copper(I) center for dioxygen activation was generated starting from flavonoid coordinating to the copper(II) site, followed by intra-molecular electron transfer to yield the copper(I) center with bound flavonol radical, which triggers dioxygen activation by the resulting copper(I) moiety [30].

2) Dioxygenase-type Dioxygen Activation and Catalysis by Redox Metal Complexes

As described earlier, the biomimetics of the monooxygenases for catalysis requires exogenous electron suppliers like $NaBH_4$ and ascorbate to facilitate dioxygen activation, which is commercially expensive and causes plenty of co-product formation. Here, in certain dioxygenases, all four electrons required for dioxygen reduction come from the substrate, which avoids the employment of an external electron supplier. Up to now, many redox metal complexes mediated dioxygen activation towards olefin oxidations have been reported. In 1985, Groves reported a $Ru^{VI}(TMP)(O)_2$ (TMP: *tetramesitylporphyrinat*) catalyzed olefin epoxidation using dioxygen as the terminal oxidant without a reducing agent added (Scheme **7**) [31]. At 25 °C in benzene, cyclooctene was selectively transformed to its epoxide, and 2 mol of epoxide were produced for each mole of dioxygen consumed. Combined with other control experiments, the authors proposed that $Ru^{VI}(TMP)(O)_2$ is the active species for olefin oxygenation. After its oxygenation, the reduced $Ru^{IV}(TMP)(O)$ proceeds through a disproportionation reaction to regenerate the active $Ru^{VI}(TMP)(O)_2$ with another $Ru^{II}(TMP)(O)$. The reduced $Ru^{II}(TMP)(O)$ next activates dioxygen to generate $Ru^{IV}(TMP)(O)$. In this catalysis, no external electron supplier was required for dioxygen activation, and both oxygen atoms from dioxygen are incorporated into olefin. With a similar concept, Che developed a chiral *trans*-$(D_4$-prophyrinato$)Ru^{VI}(O)_2$ catalyst and achieved aerobic enantioselective epoxidation of olefin in CH_2Cl_2 [32]. The proposed mechanism for this enantioselective epoxidation was similar to that of

Groves. Later, Che further explored an aerobic oxidation of terminal aryl olefin to aldehyde with Ru^{IV}(TMP)Cl$_2$ catalyst in the presence of NaHCO$_3$ [33]. In this catalysis, while the Ru complex catalyzes olefin epoxidation as well as described above, NaHCO$_3$ catalyzes the isomerization of *in situ* generated epoxide to aldehyde. Katsuki later disclosed a chiral Mn(Salen*) complex catalyzed asymmetric epoxidation with air, which provided the ee value as high as 91% [34]. Although further mechanistic information was not disclosed, it may resemble their previous photo-irradiation dioxygen activation with a similar Ru catalyst, and no reducing agent was applied in catalysis [35].

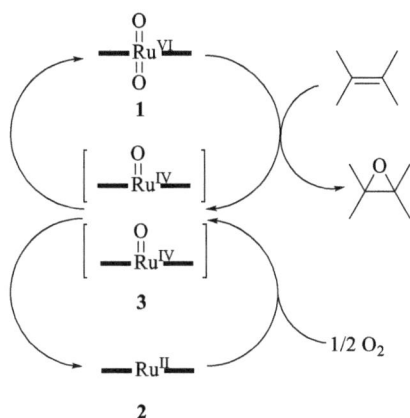

Scheme (7). Proposed catalytic mechanism for aerobic epoxidation of olefin by Ru complex.

In nature, the abundant and low toxic iron is the most preferred metal source in redox enzymes for oxygenation process, which is also central for biomimetic dioxygen activation, however, the related catalytic oxidation was not very successful until recently. In 2015, Xiao demonstrated an iron(III) complex catalyzed aerobic C=C cleavage reaction of aryl alkene, which provided aldehyde or ketone as the product (Scheme **8**) [36]. In the mechanism, the authors proposed an iron(III) based alkene ligation followed by dioxygen activation to generate an iron(IV) superoxo radical intermediate. This superoxo radical proceeds a radical cyclization with the ligated alkene, affording a five-member peroxo-metallacycle, in which the iron moiety is at the +5 state. The latter reductive elimination of this intermediate gives the dioxetane, which readily decomposes to give the carbonyl products including aldehyde and ketone, depending on the structure of the alkene substrate. The occurrences of the high valent iron(IV) superoxo and iron(V) metallocycle were attributed to the π-donation ability of the amido ligands which may stabilize the iron species at the high valence.

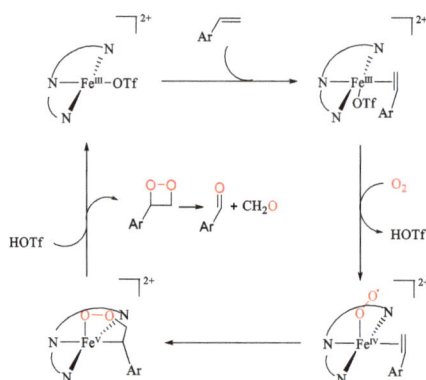

Scheme (8). Proposed catalytic mechanism for iron(III) complex catalyzed aerobic alkene C=C cleavage.

New Strategies in Dioxygen Activation and Catalysis

1) The Role of Hydrogen Bond in Enzymatic Dioxygen Activation

Up to now, even though the biological oxygenation mechanisms have been extensively studied, and biomimetic dioxygen activation has also been extensively investigated with versatile synthetic models, their applications in practical chemical oxidations are still limited. There may exist several factors which severely block the applications of biomimetic oxidation. One is that, if the oxygenation is in a monooxygenation process, such as a hydroxylation reaction, generally, the extra electron supplier is required to provide two electrons for dioxygen activation as well as those in monooxygenases. Although such a process is very popular in nature, in which the atomic economy can be controlled perfectly through a series of enzyme-mediated domino reactions, it is not controllable in a chemical plant whose functions are far less than that of a biological cell. The second comes from the dioxygen activation by synthetic redox catalyst. Generally, the most stable form of a redox metal complex in solution is not able to activate dioxygen. For example, in nature, the redox enzymes utilize an iron(II) or copper(I) to activate dioxygen, otherwise an iron(III) or copper(II) in the resting state needs to be reduced by NAD(P)H through an electron transfer chain prior to dioxygen activation. In a chemical oxidation environment, the stable iron(III) or copper(II) species are also not able to activate dioxygen efficiently except in the unusual case described by Xiao [36], and the synthetic iron(II) or copper(I) catalyst may be feasibly oxidized to its resting state, that is, iron(III) or copper(II), when exposed to air, leading to its deactivation in dioxygen activation. In certain cases as well as those in intradiol C-C cleavage dioxygenases and flavonol 2,4-dioxygeanses [29, 30], if the substrate can *in situ* reduce the iron(III) or copper(II) center to an iron(II) or copper(I) moiety through intra-molecular electron transfer upon the substrate binding, a synthetic catalyst at its resting iron(III) or copper(II) state may catalyze substrate oxidation, however, the challenge is that these reactions are not so valuable in current industrial productions. However, with the depletion of fossil resources, exploring catalysts to utilize renewable biomass as the carbon source of the chemical industry has attracted much attention. For this purpose, an introdiol dioxygenase inspired lignin oxidation with synthetic iron(III) catalyst to the aromatic ring opening products would be very attractive, because in this case, an intramolecular electron transfer from lignin based catechol to the synthetic iron(III) catalyst can trigger the dioxygen activation directly. The last one, in addition to the above mentioned challenges in biomimetic oxidations, is that the dioxygen activation efficiency of a synthetic catalyst is far less than those in native enzymes. A classic example is dioxygen binding and dioxygen activation of heme-iron in different enzymes. In Hemoglobin, the hemeiron(II) only binds and carries dioxygen, but does not activate it to superoxo radical, whereas in P450s, dioxygen is activated to the superoxo radical for catalysis [5]. Even more, in carotenoid cleavage dioxygenases, it was proposed that the dioxygen binding and activation to a superoxo radical could be reversible in catalysis [37]. Apparently, the dioxygen binding and activation can be smoothly controlled by redox enzymes, therefore, the strategy to drive the equilibrium from dioxygen binding to dioxygen activation by the synthetic catalyst is also crucial for an efficient chemical oxidation.

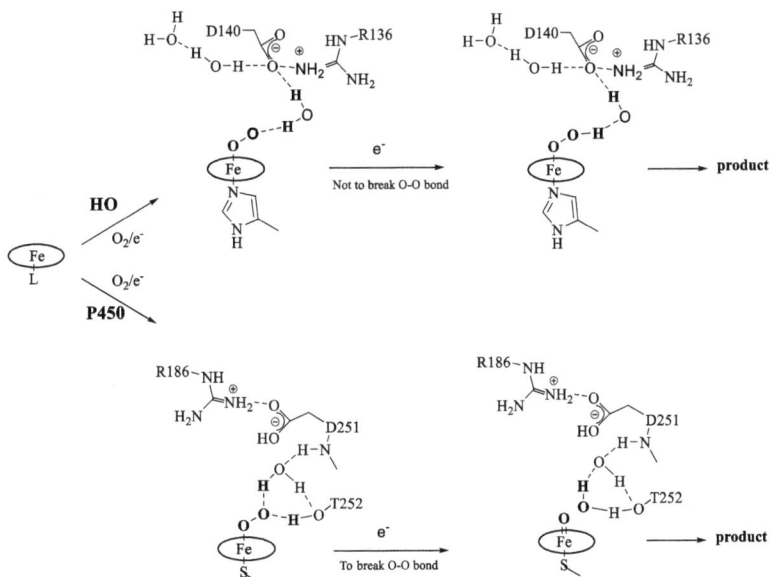

Scheme (9). The hydrogen bond network involved dioxygen activation in P450 and hemeoxygenase.

Actually, in nature, not only the first coordination sphere, for example, the heme and the axial histinate residue in P450s, of the redox metal ions plays a crucial role in dioxygen activation, the hydrogen bond network surrounding the active site also plays a significant role in stabilizing the dioxygen activation intermediate, thus driving the dioxygen activation proceeding as shown in Scheme **9** [38, 39]. However, in traditional synthetic models, only the first coordination sphere of the redox metal ions was mimicked for dioxygen activation. In this case, dioxygen may be bound to the metal ions and activated to a certain content, however, the activated oxygen species, that is, the superoxo radical, may not be stabilized due to the absence of hydrogen bond network in the synthetic model, and even more, the next hetero-cleavage of the O-O bond to generate an active oxometal species for catalysis was also not promoted, unlike those in enzymes as shown in Scheme **9**.

In 1998, Masuda demonstrated an early example of copper(II) hydroperoxide complex having bis(6-pivalamide-2-pyridylmethyl)-(2-pyridylmethyl)amine (bppa) ligand, in which the copper(II) hydroperoxide was generated from hydrogen peroxide binding rather than dioxygen activation (Scheme **10**) [40]. The presence of the intra-molecular hydrogen bond between the oxygen atom of the hydroperoxide with the NH group of the bppa ligand stabilized the copper(II) hydroperoxide, which facilitated crystal growth and made it stable in solution for more than one month at room temperature. Later, they further prepared a (μ-peroxo)dicopper(II) species through dioxygen activation with a copper(I) complex having {[6-(pivalamido)pyrid-2-yl]methyl}bis(pyrid-2-ylmethyl)amine (MPPA) ligand. The thermal stability of this dioxygen activation derived (μ-peroxo)dicopper(II) species was also attributed to the formation of the intramolecular hydrogen bond between the NH group of the ligand with the peroxide group [41].

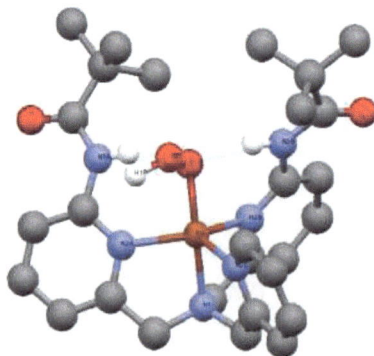

Scheme (10). Hydrogen bond stabilized hydroperoxide in synthetic copper(II) complex.

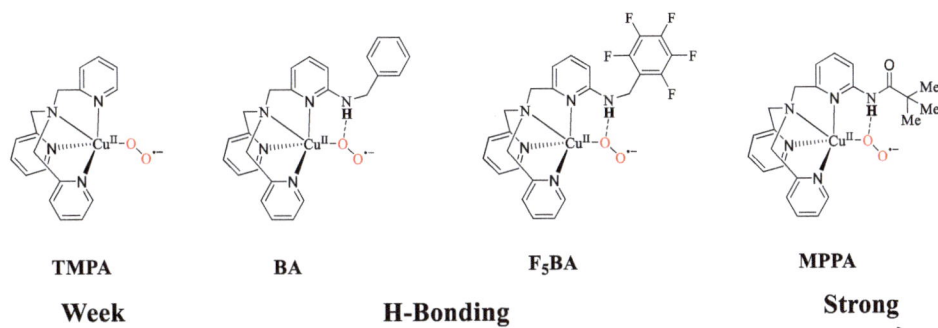

Scheme (11). Hydrogen bond stabilized superoxo radical in synthetic copper(II) complexes.

In another study, Karlin demonstrated how the intra-molecular hydrogen bond stabilizes a copper(II) superoxo radical species, which was generated through dioxygen activation with the corresponding copper(I) complex, and next affects its reactivity in hydrogen atom abstract from phenolic C-H bonds (Scheme **11**) [42]. In a series of TMPA-based copper(I) complexes (TMPA; tris(2-pyridy--methyl)amine), through ligand modifications, the ligands can provide a progressively enhanced hydrogen bond ability to stabilize the copper(II) superoxo radical, thus modifying its reactivity. The dioxygen activation by the copper(I) complexes with different TMPA-based ligands at -135 °C disclosed that the modified TMPA ligand having a strong hydrogen bond site can stabilize the generated Cu(II) superoxo radical intermediate, and a stronger hydrogen bond provides better stability of the Cu(II) superoxo radical species. Remarkably, the hydrogen bond bound Cu(II) superoxo radical reacts more efficiently in hydrogen abstraction from phenolic O-H bond, thus resembling those hydrogen bond network modulated dioxygen activation and reactivity in redox enzymes [38, 39]. Using the biotin-streptavidin (Sav) technology, Borovik developed the artificial copper proteins to stabilize a Cu^{II}-OOH complex through a hydrogen bond network inside the artificial protein, where the Cu^{II}-OOH moiety was generated by treating Cu^{II} complexes with H_2O_2 [43]. Through correlating the changes in the reactivity of the Cu^{II}-OOH species with modulation of the hydrogen bonds, the authors highlighted the crucial role of the hydrogen bonds in regulating the stability and reactivity of the Cu^{II}-OOH species, that is, the single hydrogen bond linked to the proximal oxygen atom of the Cu^{II}-OOH species can stabilize the Cu^{II}-OOH species within the protein host, while the hydrogen bond at the distal oxygen atom activates the Cu^{II}-OOH species to oxidize the substrate like 4-chlorobenzylamine.

2) Lewis Acid Promoted Dioxygen Activation and Catalysis

Above pioneering works in dioxygen activation with synthetic models have highlighted that not only the first coordination sphere of the redox metal ions plays a crucial role in dioxygen activation, but the surrounding hydrogen bond network may also play the key role in stabilizing the *in-situ* activated superoxo radical intermediate, and further modify its reactivity in oxidations, thus driving the catalysis forward. Therefore, both the first coordination sphere of the metal ions and the surrounding hydrogen bond network are essential for achieving an efficient dioxygen activation and catalysis. However, the challenge is that the hydrogen bond network in a chemical reaction environment is more difficult to control than those in the proteins. Accordingly, alternative strategy may need to be explored to play the role as well as the hydrogen bond network in redox enzymes to drive the dioxygen activation in a synthetic model.

One significant role of the hydrogen bond network in dioxygen activation is that the electrostatic interaction between the proton and the superoxo radical can stabilize it, and drive the equilibrium of dioxygen activation forward to the superoxo radical side, and next drive the O-O bond cleavage to generate the active oxometal intermediate for oxidation in enzymes [38, 39]. In addition, the enhanced electrophilic properties of the superoxo radical by binding a proton through hydrogen bond may also improve its reactivity for oxidation as disclosed in synthetic models [42]. Resembling the hydrogen bond interaction, a Lewis acid may also interact with the superoxo radical through electrostatic interaction (Brönsted acid *vs* Lewis acid), and the choice of the Lewis acid would be more flexible than Brönsted acid. An interesting fact is that, in investigating the electron transfer between electron donor (D) and acceptor (A), Fukuzumi found that, the presence of a metal ion as Lewis acid can drive the electron transfer from the donor to acceptor, because it stabilizes the $A^{\cdot-}$ anion (Scheme **12**) [44]. Similarly, dioxygen activation by a redox metal ion to generate the superoxo radical is also one category of electron transfer between electron donor and acceptor. After the dioxygen activation, the generated superoxo radical intermediate is also an anion; accordingly, a positively charged Lewis acid may interact with the superoxo radical anion and stabilize it through electrostatic interaction, resulting in the dioxygen activation smoothly.

$$\text{D} + \text{A} \xrightarrow{\text{ET}} \text{D}^{\bullet+} + \text{A}^{\bullet-}$$

$$+ \text{LA}^{n+} \Updownarrow \qquad\qquad \text{Concerted MCET} \qquad\qquad \downarrow + \text{LA}^{n+}$$

$$\text{D} + \text{A}{-}\text{LA}^{n+} \xrightarrow{\text{ET}} \text{D}^{\bullet+} + \text{A}^{\bullet-}{-}\text{LA}^{n+}$$

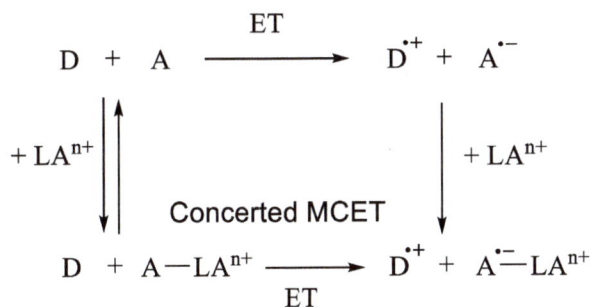

Scheme (12). Lewis acid promoted electron transfer between electron donor and acceptor.

Based on this concept, in 2017, Yin explored a Lewis acid promoted dioxygen activation by vanadium(IV) complex for catalytic hydrogen abstraction reaction [45]. Using 1,4-cyclohexadiene as a testing substrate, in CH_3CN, the presence of Lewis acid substantially accelerated $V^{IV}(TPA)$ complex (TPA: tris-[(--pyridy)methyl]amine) catalyzed hydrogen atom abstraction from 1,4-cyclohexadiene to benzene with O_2 balloon as the oxygen source at 40 °C, and the catalytic efficiency was highly Lewis acidity dependent of the added non-redox metal ions. Mechanistic studies revealed that the presence of Lewis acid can stabilize the $V^V(TPA)$-$O_2^{\cdot-}$ superoxo species, which was generated through dioxygen activation by $V^{IV}(TPA)$ complex, thus driving the equilibrium of dioxygen activation towards the formation of the $V^V(TPA)$-$O_2^{\cdot-}$ superoxo species, and next accelerating hydrogen abstraction from the substrate (Scheme **13**). In the absence of Lewis acid, such a hydrogen abstraction reaction was very sluggish. This is an early example of Lewis acid promoted dioxygen activation by redox metal ions towards catalytic oxidation; although it is an oxidase-type biomimetics of dioxygen activation, it has already illustrated a novel strategy in mimicking dioxygen activation towards efficient catalysis.

(Scheme 13) contd.....

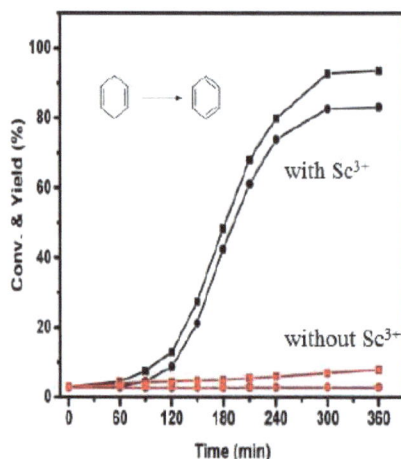

Scheme (13). Lewis acid promoted dioxygen activation by vanadium(IV) complex towards catalytic hydrogen abstraction.

CONCLUSIVE MARKS AND PERSPECTIVE

Because of the complexity and high synergy of enzymes in electron transfer, dioxygen activation and next oxidation, event that the mechanisms for enzymatic dioxygen activation and oxidations have been much clearly elucidated than before, the progress in biomimetic oxidations with simple synthetic metal complex is far less than successful for industrial applications, and most of the chemical oxidations in the industry with dioxygen are still performed at elevated temperature with low selectivity. Typically for those monooxygenases, the requirement of extra electron supplier has badly prevented its biomimetic applications in industry. After its early biomimetics with $NaBH_4$, ascorbic acid, *etc.*, as the reducing agent, its further tests for industrial applications have almost disappeared. The viewable light for biomimetic oxidation may come from dioxygenase-type oxidation, which does not need an extra electron supplier in certain cases, and some biomimetic examples in olefin epoxidation and C=C bond cleavage have evidenced its validity. To improve the efficiency in dioxygen activation, a strategy of Lewis acid promoted dioxygen activation by redox metal complexes was also proposed recently, and demonstrated a high efficiency in catalytic hydrogen atom abstraction reaction. Possibly, introducing Lewis acid promoted dioxygen activation strategy into the dioxygenase-type biomimetics may have the opportunity to improve its catalytic efficiency, thus leading to an industrial application. However, biomimetic dioxygen activation toward efficient catalysis is still at its infant stage, and plenty of works are still underway in improving the efficiency of dioxygen activation and in controlling the reactivity and selectivity of these active oxygen species in catalysis.

CONSENT FOR PUBLICATION

Not Applicable.

CONFLICT OF INTEREST

The authors declare no conflict of interest, financial or otherwise.

ACKNOWLEDGEMENTS

Declared none.

REFERENCES

[1]   Bart, J.C.J.; Cavallaro, S. Transiting from adipic acid to bioadipic acid. 1, petroleum-based processes. *Ind. Eng. Chem. Res.,* **2015**, *54*(1), 1-46.
        [http://dx.doi.org/10.1021/ie5020734]

[2]   Steeman, J.W.M.; Kaarsemaker, S.; Hoftijzer, P.J. Pilot plant study of the oxidation of cyclohexane with air under pressure. *Chem. Eng. Sci.,* **1961**, *14*(1), 139-150.
        [http://dx.doi.org/10.1016/0009-2509(61)85066-5]

[3]   Special issue, 60 years of dioxygen activation. *J. Biol. Inorg. Chem.,* **2017**, *22*(2-3)

[4]   Sahu, S.; Goldberg, D.P. Activation of dioxygen by iron and manganese complexes: A heme and nonheme perspective. *J. Am. Chem. Soc.,* **2016**, *138*(36), 11410-11428.
        [http://dx.doi.org/10.1021/jacs.6b05251] [PMID: 27576170]

[5]   Sono, M.; Roach, M.P.; Coulter, E.D.; Dawson, J.H. Heme-Containing Oxygenases. *Chem. Rev.,* **1996**, *96*(7), 2841-2888.
        [http://dx.doi.org/10.1021/cr9500500] [PMID: 11848843]

[6]   Denisov, I.G.; Makris, T.M.; Sligar, S.G.; Schlichting, I. Structure and chemistry of cytochrome P450. *Chem. Rev.,* **2005**, *105*(6), 2253-2277.
        [http://dx.doi.org/10.1021/cr0307143] [PMID: 15941214]

[7]   Groves, J.T. Key elements of the chemistry of cytochrome P-450, the oxygen rebound mechanism. *J. Chem. Educ.,* **1985**, *62*(11), 928-931.
        [http://dx.doi.org/10.1021/ed062p928]

[8]   Ross, M.O.; Rosenzweig, A.C. A tale of two methane monooxygenases. *Eur. J. Biochem.,* **2017**, *22*(2-3), 307-319.
        [http://dx.doi.org/10.1007/s00775-016-1419-y] [PMID: 27878395]

[9]   Baik, M-H.; Newcomb, M.; Friesner, R.A.; Lippard, S.J. Mechanistic studies on the hydroxylation of methane by methane monooxygenase. *Chem. Rev.,* **2003**, *103*(6), 2385-2419.
        [http://dx.doi.org/10.1021/cr950244f] [PMID: 12797835]

[10]   Cutsail, G.E., III; Banerjee, R.; Zhou, A.; Que, L., Jr; Lipscomb, J.D.; DeBeer, S. High-resolution extended x-ray absorption fine structure analysis provides evidence for a longer Fe⋯Fe distance in the Q intermediate of methane monooxygenase. *J. Am. Chem. Soc.,* **2018**, *140*(48), 16807-16820.
        [http://dx.doi.org/10.1021/jacs.8b10313] [PMID: 30398343]

[11]   Lieberman, R.L.; Rosenzweig, A.C. Biological methane oxidation: regulation, biochemistry, and active site structure of particulate methane monooxygenase. *Crit. Rev. Biochem. Mol. Biol.,* **2004**, *39*(3), 147-164.
        [http://dx.doi.org/10.1080/10409230490475507] [PMID: 15596549]

[12]   Himes, R.A.; Barnese, K.; Karlin, K.D. One is lonely and three is a crowd: two coppers are for

methane oxidation. *Angew. Chem. Int. Ed. Engl.*, **2010**, *49*(38), 6714-6716.
[http://dx.doi.org/10.1002/anie.201003403] [PMID: 20672276]

[13]   Tabushi, I.; Koga, N. P-450 type oxygen activation by porphyrin-manganese complex. *J. Am. Chem. Soc.*, **1979**, *101*(21), 6456-6458.
[http://dx.doi.org/10.1021/ja00515a063]

[14]   Mansuy, D.; Fontecave, M.; Bartoli, J-F. Mono-oxygenase-like dioxygen activation leading to alkane hydroxylation and olefin epoxidation by an Mn^{III}(porphyrin)-ascorbate biphasic system. *J. Chem. Soc. Chem. Commun.*, **1983**, *6*(6), 253-254.
[http://dx.doi.org/10.1039/C39830000253]

[15]   Murahashi, S-I.; Naota, T.; Komiya, N. Metalloporphyrin-catalyzed oxidation of alkanes with molecular oxygen in the presence of acetaldehyde. *Tetrahedron Lett.*, **1995**, *36*(44), 8059-8062.
[http://dx.doi.org/10.1016/0040-4039(95)01708-P]

[16]   Brégeault, J-M. Transition-metal complexes for liquid-phase catalytic oxidation: some aspects of industrial reactions and of emerging technologies. *Dalton Trans.*, **2003**, (17), 3289-3302.
[http://dx.doi.org/10.1039/B303073N]

[17]   Siegbahn, P.E.M.; Crabtree, R.H. Mechanism of C-H activation by diiron methane monooxygenases: Quantum chemical studies. *J. Am. Chem. Soc.*, **1997**, *119*(13), 3103-3113.
[http://dx.doi.org/10.1021/ja963939m]

[18]   Xue, G.; De Hont, R.; Münck, E.; Que, L., Jr Million-fold activation of the [Fe(2)(micro-O)(2)] diamond core for C-H bond cleavage. *Nat. Chem.*, **2010**, *2*(5), 400-405.
[http://dx.doi.org/10.1038/nchem.586] [PMID: 20414242]

[19]   Xue, G.; Pokutsa, A.; Que, L., Jr Substrate-triggered activation of a synthetic [Fe$_2$(μ-O)$_2$] diamond core for C-H bond cleavage. *J. Am. Chem. Soc.*, **2011**, *133*(41), 16657-16667.
[http://dx.doi.org/10.1021/ja207131g] [PMID: 21899336]

[20]   Chen, Z.; Yang, L.; Choe, C.; Lv, Z.; Yin, G. Non-redox metal ion promoted oxygen transfer by a non-heme manganese catalyst. *Chem. Commun. (Camb.)*, **2015**, *51*(10), 1874-1877.
[http://dx.doi.org/10.1039/C4CC07981G] [PMID: 25525748]

[21]   Choe, C.; Yang, L.; Lv, Z.; Mo, W.; Chen, Z.; Li, G.; Yin, G. Redox-inactive metal ions promoted the catalytic reactivity of non-heme manganese complexes towards oxygen atom transfer. *Dalton Trans.*, **2015**, *44*(19), 9182-9192.
[http://dx.doi.org/10.1039/C4DT03993A] [PMID: 25904197]

[22]   Zhang, J.; Wei, W-J.; Lu, X.; Yang, H.; Chen, Z.; Liao, R-Z.; Yin, G. Nonredox metal ions promoted olefin epoxidation by iron(II) complexes with H$_2$O$_2$: DFT calculations reveal multiple channels for oxygen transfer. *Inorg. Chem.*, **2017**, *56*(24), 15138-15149.
[http://dx.doi.org/10.1021/acs.inorgchem.7b02463] [PMID: 29182327]

[23]   Liu, Y.; Lau, T.C. T,-C, Lau, Activation of metal oxo and nitrido complexes by lewis acid. *J. Am. Chem. Soc.*, **2019**, *141*(9), 3755-3766.
[http://dx.doi.org/10.1021/jacs.8b13100] [PMID: 30707842]

[24]   Devi, T.; Lee, Y-M.; Nam, W.; Fukuzumi, S. Metal ion-coupled electron-transfer reactions of metal-oxygen complexes. *Coord. Chem. Rev.*, **2020**, *410*213219
[http://dx.doi.org/10.1016/j.ccr.2020.213219]

[25]   Guo, H.; Chen, Z.; Mei, F.; Zhu, D.; Xiong, H.; Yin, G. Redox inactive metal ion promoted C-H activation of benzene to phenol with Pd$^{(II)}$(bpym): demonstrating new strategies in catalyst designs. *Chem. Asian J.*, **2013**, *8*(5), 888-891.
[http://dx.doi.org/10.1002/asia.201300003] [PMID: 23401395]

[26]   Qin, S.; Dong, L.; Chen, Z.; Zhang, S.; Yin, G. Non-redox metal ions can promote Wacker-type oxidations even better than copper(II): a new opportunity in catalyst design. *Dalton Trans.*, **2015**, *44*(40), 17508-17515.

[http://dx.doi.org/10.1039/C5DT02612A] [PMID: 26390300]

[27] Xue, J-W.; Zeng, M.; Zhang, S.; Chen, Z.; Yin, G. Lewis acid promoted aerobic oxidative coupling of thiols with phosphonates by simple nickel(II) catalyst: Substrate scope and mechanistic studies. *J. Org. Chem.*, **2019**, *84*(7), 4179-4190.
[http://dx.doi.org/10.1021/acs.joc.9b00194] [PMID: 30870591]

[28] Chung, L.W.; Li, X.; Sugimoto, H.; Shiro, Y.; Morokuma, K. ONIOM study on a missing piece in our understanding of heme chemistry: bacterial tryptophan 2,3-dioxygenase with dual oxidants. *J. Am. Chem. Soc.*, **2010**, *132*(34), 11993-12005.
[http://dx.doi.org/10.1021/ja103530v] [PMID: 20698527]

[29] Wang, Y.; Li, J.; Liu, A. Oxygen activation by mononuclear nonheme iron dioxygenases involved in the degradation of aromatics. *Eur. J. Biochem.*, **2017**, *22*(2-3), 395-405.
[http://dx.doi.org/10.1007/s00775-017-1436-5] [PMID: 28084551]

[30] Pap, J.S.; Kaizer, J.; Speier, G. Model systems for the CO-releasing flavonol 2,4-dioxygenase enzyme. *Coord. Chem. Rev.*, **2010**, *254*(7-8), 781-793.
[http://dx.doi.org/10.1016/j.ccr.2009.11.009]

[31] Groves, J.T.; Quinn, R. Aerobic epoxidation of olefins with ruthenium porphyrin catalysts. *J. Am. Chem. Soc.*, **1985**, *107*(20), 5790-5792.
[http://dx.doi.org/10.1021/ja00306a029]

[32] Lai, T-S.; Zhang, R.; Cheung, K-K.; Kwong, H-L.; Che, C-M. Aerobic enantioselective alkene epoxidation by a chiral trans-dioxo(D4-porphyrinato)ruthenium(vi) complex. *Chem. Commun. (Camb.)*, **1998**, (15), 1583-1584.
[http://dx.doi.org/10.1039/a802009d]

[33] Jiang, G.; Chen, J.; Thu, H-Y.; Huang, J-S.; Zhu, N.; Che, C-M. Ruthenium porphyrin-catalyzed aerobic oxidation of terminal aryl alkenes to aldehydes by a tandem epoxidation-isomerization pathway. *Angew. Chem. Int. Ed. Engl.*, **2008**, *47*(35), 6638-6642.
[http://dx.doi.org/10.1002/anie.200801500] [PMID: 18651688]

[34] Koya, S.; Nishioka, Y.; Mizoguchi, H.; Uchida, T.; Katsuki, T. Asymmetric epoxidation of conjugated olefins with dioxygen. *Angew. Chem. Int. Ed. Engl.*, **2012**, *51*(33), 8243-8246.
[http://dx.doi.org/10.1002/anie.201201848] [PMID: 22821760]

[35] Tanaka, H.; Nishikawa, H.; Uchida, T.; Katsuki, T. Photopromoted Ru-catalyzed asymmetric aerobic sulfide oxidation and epoxidation using water as a proton transfer mediator. *J. Am. Chem. Soc.*, **2010**, *132*(34), 12034-12041.
[http://dx.doi.org/10.1021/ja104184r] [PMID: 20701287]

[36] Gonzalez-de-Castro, A.; Xiao, J. Green and efficient: Iron-catalyzed selective oxidation of olefins to carbonyls with O_2. *J. Am. Chem. Soc.*, **2015**, *137*(25), 8206-8218.
[http://dx.doi.org/10.1021/jacs.5b03956] [PMID: 26027938]

[37] Harrison, P.J.; Bugg, T.D.H. Enzymology of the carotenoid cleavage dioxygenases: reaction mechanisms, inhibition and biochemical roles. *Arch. Biochem. Biophys.*, **2014**, *544*, 105-111.
[http://dx.doi.org/10.1016/j.abb.2013.10.005] [PMID: 24144525]

[38] Momenteau, M.; Reed, C.A. Synthetic heme dioxygen complexes. *Chem. Rev.*, **1994**, *94*(3), 659-698.
[http://dx.doi.org/10.1021/cr00027a006]

[39] Fujii, H.; Zhang, X.; Tomita, T.; Ikeda-Saito, M.; Yoshida, T. A role for highly conserved carboxylate, aspartate-140, in oxygen activation and heme degradation by heme oxygenase-1. *J. Am. Chem. Soc.*, **2001**, *123*(27), 6475-6484.
[http://dx.doi.org/10.1021/ja010490a] [PMID: 11439033]

[40] Wada, A.; Harata, M.; Hasegawa, K.; Jitsukawa, K.; Masuda, H.; Mukai, M.; Kitagawa, T.; Einaga, H. Structural and spectroscopic characterization of a mononuclear hydroperoxo – copper(II) complex with tripodal pyridylamine ligands. *Angew. Chem. Int. Ed. Engl.*, **1998**, *37*(6), 798-799.

[http://dx.doi.org/10.1002/(SICI)1521-3773(19980403)37:6<798::AID-ANIE798>3.0.CO;2-3]
[PMID: 29711401]

[41]   Yamaguchi, S.; Wada, A.; Funahashi, Y.; Nagatomo, S.; Kitagawa, T.; Jitsukawa, K.; Masuda, H. Thermal stability and absorption spectroscopic behavior of (μ-peroxo)dicopper complexes regulated with intramolecular hydrogen bonding interactions. *Eur. J. Inorg. Chem.,* **2003**, *2003*(24), 4378-4386.
[http://dx.doi.org/10.1002/ejic.200300178]

[42]   Bhadra, M.; Lee, J.Y.C.; Cowley, R.E.; Kim, S.; Siegler, M.A.; Solomon, E.I.; Karlin, K.D. Intramolecular hydrogen bonding enhances stability and reactivity of mononuclear cupric superoxide complexes. *J. Am. Chem. Soc.,* **2018**, *140*(29), 9042-9045.
[http://dx.doi.org/10.1021/jacs.8b04671] [PMID: 29957998]

[43]   Mann, S.I.; Heinisch, T.; Ward, T.R.; Borovik, A.S. Peroxide activation regulated by hydrogen bonds within artificial Cu proteins. *J. Am. Chem. Soc.,* **2017**, *139*(48), 17289-17292.
[http://dx.doi.org/10.1021/jacs.7b10452] [PMID: 29117678]

[44]   Fukuzumi, S.; Ohkubo, K.; Morimoto, Y. Mechanisms of metal ion-coupled electron transfer. *Phys. Chem. Chem. Phys.,* **2012**, *14*(24), 8472-8484.
[http://dx.doi.org/10.1039/c2cp40459a] [PMID: 22596095]

[45]   Zhang, J.; Yang, H.; Sun, T.; Chen, Z.; Yin, G. Nonredox metal-ions-enhanced dioxygen activation by oxidovanadium(IV) complexes toward hydrogen atom abstraction. *Inorg. Chem.,* **2017**, *56*(2), 834-844.
[http://dx.doi.org/10.1021/acs.inorgchem.6b02277] [PMID: 28055194]

CHAPTER 3

Highlights of Oxygen Atom Transfer Reactions Catalysed by Nickel Complexes

Anjana Rajeev[1] and **Muniyandi Sankaralingam**[1,*]

[1] *Bioinspired & Biomimetic Inorganic Chemistry Laboratory, Department of Chemistry, National Institute of Technology Calicut, Kozhikode, Kerala-673601, India*

Abstract: Oxygen atom transfer (OAT) reactions catalyzed by metal complexes have been a subject of intensive research over the century, owing to the prevalent involvement of OAT in organic transformations mediated by several biologically important enzymes such as methane monooxygenases, cytochrome P450, etc as well as in synthetic chemical reactions. In biomimetic model chemistry, iron and manganese complexes are the most exploited catalysts in the realm of oxygen atom transfer reactions and many of these metal complexes produce very short-lived reactive metal-oxygen intermediates during the catalytic reactions. Characterization of such reactive intermediates of numerous heme and non-heme iron and manganese complexes and comparing them with their natural enzyme analogous have emerged as a promising approach toward understanding several intricate enzymatic mechanisms. Considerable research advancements in the studies of OAT reactions involving late transition metal complexes such as cobalt, nickel, and copper have also been recognized in the past few years. In this account, various reports have been published, demonstrating catalytic oxidation of organic substrates by the active nickel-oxygen species generated either *via* heterolysis or homolysis of O-O bond of oxidant bound nickel complexes. This book chapter aims at a comprehensive summary of noteworthy attempts contributed towards nickel catalyzed OAT reactions and various implicated or well-characterized nickel-oxygen active intermediates. The effect of stereoelectronic properties of ligand architecture on catalytic efficiency and various characterization techniques used to identify the catalytically active nickel-oxygen species are also discussed.

Keywords: Oxo atom transfer reaction catalyzed by nickel complexes, Hydrogen atom abstraction reaction catalyzed by nickel complexes.

INTRODUCTION

In nature, enzymes are the most complicated yet fundamental molecules. Apprehending the intricate catalytic processes involved in enzymatic reactions

* **Corresponding author Muniyandi Sankaralingam:** Bioinspired & Biomimetic Inorganic Chemistry Lab, Department of Chemistry, National Institute of Technology Calicut, Kozhikode, Kerala-673601, India; Tel: +914952285325; E-mails: msankaralingam@nitc.ac.in, sankarjan06@gmail.com

Robert Bakhtchadjian (Ed.)

has remained a challenge until the emergence of biomimetic chemistry by bioinorganic chemists. Biomimetic chemistry involves the design, synthesis, and reactivity studies of model systems of biological molecules and it has become an effective approach to comprehend complicated enzymatic reactions [1 - 10]. Rigorous research on various model complexes of non-heme and heme enzymes has paved the way for reinforcement of our understanding of the detailed mechanism and the process of dioxygen activation in various enzymes [11 - 22]. Iron is one of the main protagonists in the field of oxo atom transfer reactions and cytochrome P450 (CYP 450) is the most celebrated and studied enzyme in this area. Identification of the involvement of a high-valent iron(IV)-oxo porphyrin cation radical species (Compound I) generated *via* putative heterolytic cleavage of O-O bond in the catalytic cycle of CYP 450s has attracted great research interest among biomimetic chemists [23]. Intensive research efforts to examine the chemical and physical properties of compound I to compare this intermediate with other such enzymatic analogous have always been a matter of interest. Organic transformation reactions especially oxidation reactions such as epoxidation, sulfoxidation, hydroxylations, carbonylation, *etc.* have fascinated a lot of researchers because of their biological and industrial importance. Epoxides are useful intermediates in the production of polyurethane, polyamides, polyesters, resins, and in bioinspired oxidation to produce drug metabolites. Sulfur-containing compounds are well known for their antimicrobial activity and hydroxylated compounds are widely used in petrochemical industry [24]. The first synthetic iron(III) porphyrin complex catalyst for olefin epoxidation and alkane hydroxylation was reported in 1979 by Groves *et al.* [25] and, iron porphyrin chemistry has been a subject of study since then. Various metalloporphyrins models of CYP450s capable of catalyzing the oxidation of hydrocarbons, alcohols, sulfides, and olefins have been reported over the past three decades [26]. Rahimi and co-workers demonstrated a CYP450 model reaction by a Cu(II) meso-tetraphenyl porphyrin in the oxidation of benzyl alcohols to corresponding carbonyl compounds [27]. Nam *et al.* reported the reaction pathways of O-O bond cleavage in hydrogen peroxide and *tert*-alkyl hydroperoxides using iron(III) porphyrin complexes. On the one hand, electron rich porphyrins and hydroperoxide having electron-releasing substituents tend to favor the homolytic cleavage of O-O bond. On the other hand, heterolytic O-O bond cleavage is facilitated by electron-deficient porphyrins and electron-withdrawing group substituted hydroperoxide. Moreover, analysis of the products obtained after iron porphyrin catalyzed the epoxidation reaction of olefins in the presence of hydrogen peroxide or alkyl hydroperoxide can be used to distinguish the mode of O-O bond cleavage. The heterolytic cleavage results showed the formation of stereospecific epoxide with high yields, whereas homolytic cleavage afforded less stereospecific epoxide in lower yield [28]. Very recently, Nam *et al.* reported the remarkable reactivities of Mn(III)-iodosylarene porphyrins in C-H bond activation and oxygen atom transfer reactions. Interestingly, unlike the iron-oxo porphyrins, the reactivities of these complexes are found to be independent of the electronic nature of the porphyrin ligands [29].

Despite being a trace element, nature prefers nickel center as an active site in a few enzymes such as glyoxylase I, nickel superoxide dismutase, urease, NiFe hydrogenase, CO dehydrogenase, acetyl-CoA synthase, and, methyl-CoM reductase [30]. After the discovery of the first-ever nickel-containing metalloenzyme urease, many researchers have been interested to uncover the involvement of nickel in biologically important molecules. Flexibility in its coordination and redox chemistry draws a lot of research interest towards nickel. The presence of nickel center in a few biologically available enzymes such as Ni-superoxide dismutase and quercetin 2,4-dioxygenase that are involved in oxidation reactions has gained considerable research interest in the area of nickel mediated oxidation chemistry [31]. However, when compared to iron and manganese, reports on oxidation reactions involving nickel chemistry are sparse. A number of comprehensive and high-level reviews and book chapters have been published over the years notably concerning the use of iron, manganese, and copper complexes as bioinspired catalysts for the variety of oxidation reactions in biomimetic chemistry [32 - 65]. However, less emphasis has been put on the oxo transfer reactions using nickel complexes but related approaches are developing fast [1]. So we took an opportunity and advantage to summarize the nickel-mediated oxidation chemistry.

Oxygen Atom Transfer Reactions

Oxygen atom transfer reactions such as sulfoxidation, phosphine oxidation, and epoxidation reactions are very important and the resulting products are very useful in many industrial processes. Long ago in 1969, Otsuka *et al.* demonstrated the oxygenation of alkyl isocyanides and triphenyl phosphines by a Ni(II)-peroxo complex derived from the oxygenation of zero-valent nickel isocyanide complex, $Ni(RCN)_4$ (R=*tert*-butyl or cyclohexyl). After that, a few articles on CO fixation by a series of nickel complexes such as $Ni(NO_2)_2(PMe_3)_2$ (**1**), $Ni(NO_2)_2(PEt_3)_2$ (**2**), $Ni(NO_2)_2(DPPE)$ (**3**) (where, PMe_3, trimethylphosphine, PEt_3, trimethylphosphine, and DPPE, $Ph_2PCH_2CH_2PPh_2$) have been reported. During the reaction of these complexes with CO to form CO_2, a pentacoordinated intermediate is likely to be involved. The source of oxygen atom was found to be NO_2 ligand as evident from the ^{18}O-labeling experiment. Enhanced reactivity of **1** compared to **2** also supported the notion of involvement of a pentacoordinated intermediate, $Ni(CO)(NO_2)_2(PR_3)_2$ owing to the ability of more basic PMe_3 to stabilize pentacoordinate monocarbonyl complex [66]. In 2004, Riordan *et al.* identified a side-on nickel-dioxygen intermediate prior to the generation of the bis((μ-oxo)dinickel(III) complex during the reaction of a Ni(I) complex, [PhTt^{Ad}]Ni(CO) (PhTt^{Ad}, phenyl-tris((1-adamantylthio)methyl)borate) and dioxygen. Spectroscopic studies, DFT calculations, and isotopic distribution in ESI-MS data of the dioxygen adduct suggested the formation of a side-on dioxygen adduct which was again confirmed by EXAFS studies. For instance, Rhombic signals obtained in the EPR spectrum of active oxygen intermediate were found to have three *g* values of 2.24, 2.19, and 2.01. The absence of a short M-M vector in EXAFS data ascertained the mononuclear nature of the dioxygen adduct. The side-on adduct with a square pyramidal geometry was understood to be a Ni(II)-superoxo complex with a high spin Ni(II) center antiferromagnetically coupled to the superoxide radical to possess d_z^2 ground state electronic configuration. Interestingly, this superoxide species was found to be efficient in the transfer of an oxygen atom to triphenylphosphine (PPh$_3$) and dioxygen to nitric oxide [67]. In 2008, Driess *et al.* reported the crystallographically characterized first superoxo nickel complex **4a** (the chemical structure is shown in Fig. (**1a**) and its oxygen atom transfer reactivity towards PPh$_3$. This square planar Ni(II) complex with an unpaired electron sitting on the superoxide ligand was obtained from the dry O$_2$ exposure of precursor [(Ni^I(β-diketiminato-(toluene))] complex and characterized using IR and X-band EPR spectroscopies with the backing of DFT studies. Interestingly, oxygenation of PPh$_3$ by employing **4a** afforded a paramagnetic, dinuclear {(Ni^{II})$_2$(μ-OH)$_2$} complex (**4b**) (the chemical structure is shown in Fig. (**1b**) having planar and tetrahedral Ni centers along with triphenylphosphine oxide (O=PPh$_3$). Even though insight into the detailed mechanism is not provided, they proposed that the presence of a nickel-oxo species capable of scavenging hydrogen and a subsequent dimerization is likely to be involved during the formation of **4b** [68]. In a subsequent paper, the author explored the reactivity of **4a** towards various types of substrates such as alkanes, alkenes, sulfides, etc and found that it is not capable of transferring an oxygen atom to these substrates but oxidizes O-H (2,6-di-*tert*-butylphenol (2,6-DTBP)) and N-H (1,2-diphenyl-hydrazine) bonds readily. ^{18}O-labeled studies revealed that the oxygen atom incorporated in the oxidized product of 2,6,-DTBP is from **4a**, and thus this complex is acting as an oxygenating reagent. As proposed in the case of PPh$_3$ oxidation, this oxygenation is also likely to be proceeding *via* the formation of a Ni^{III}-oxo species [69]. In 2015, Company *et al.* reported the trapping of a Ni^{III}-O˙ species generated *via* heterolytic O-O bond cleavage of *meta*-chloroperbenzoic acid by using a square planar Ni(II) complex (**5**) of a dianionic macrocyclic N4 ligand, L5 (Fig. **2**) below 250 K. Enhanced reactivity of Ni^{III}-oxyl species towards the oxidation of various organic substrates such as olefins, sulfides, and activated methylene C-H bonds is presumably due to the involvement of Ni-oxygen species which in turn point towards the influence of fine-tuning of ligand architectures on the reactivity of complexes. The negative value of the reaction constant $\rho = -0.86$ obtained after employing a series of *para*-substituted thioanisoles and styrenes as substrates indicated the electrophilic character of Ni^{III}-oxyl species. The +3 oxidation state of the transition state was ascertained by intense pre-edge $1s{\rightarrow}3d$ transition around 8333.5 eV at the metal K-edge of XAS study and higher rising-edge energy (1.5-2 eV higher than Ni^{II} species).

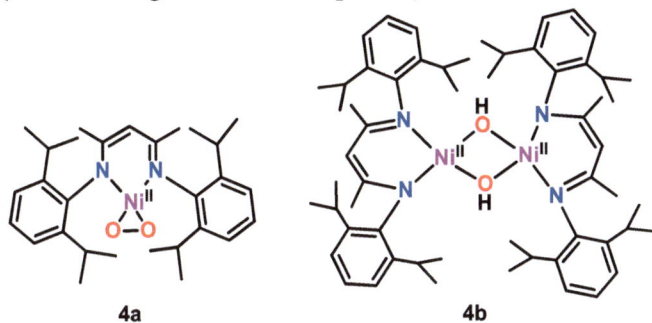

4a   **4b**

Fig. (1). The chemical structures of **4a** and **4b** [68].

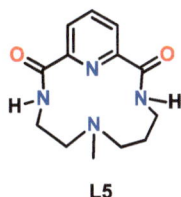

L5

Fig. (2). The structure of L5 [70].

Kinetic and thermodynamic feasibility for the formation of different plausible catalytic intermediates have been probed using DFT studies and that suggested the formation of a high-valent nickel species through O-O heterolysis (9 kcal mol^{-1}) [70].

Oxidation Involving Hydrogen Atom Abstraction Reaction

Schröder, and Schwarz, *et al.* published a series of articles based on the reactivity of bare first-row transition metal-oxide species, MO^+ in gas phase towards methane oxidation among which NiO^+ species was found to be the most promising oxidant in terms of efficiency and selectivity [71 - 74]. Later in 2000, Yoshizawa *et al.* discussed the energetics and reaction pathway behind methane to methanol conversion by first-row transition metal-oxide species by employing density functional theory (DFT) calculations. Based on this study, two feasible spin inversion taking place during the entire reaction is proposed to be responsible for better efficiency of FeO^+, NiO^+ and CuO^+ species in methane conversion when compared to less efficient conversion mediated by the early transition metal-oxide species with a single crossing point [75]. In 2001, Itoh, Kitagawa, Fukuzumi and their co-workers reported the generation of a series of bis(μ-oxo)dinickel(III) complexes of mono and dinucleating bis(pyridylalkyl)amine ligands (L6^X and L7-L10 respectively) (Fig. **3**) by the reaction of corresponding [Ni^{II}(μ-OR)$_2$ Ni^{II}] (R= H or OMe) complexes with an equimolar amount of hydrogen peroxide in acetone at low temperature. The bis(μ-oxo)dinickel(III) complexes exhibited a distinctive UV-vis absorption band at 410 nm and a resonance Raman (rRaman) band at 600-610 cm^{-1}. Based on the investigation of the formation of [Ni^{III}(μ-O)$_2$ Ni^{III}] complexes, it is proposed that a rate-determining isomerization of [Ni$^{II}\mu$-OR)$_2$Ni^{II}] complex preceding H$_2$O$_2$ attack is likely to be involved in the process which is followed by the formation of a ($\mu-\eta^2$:η^2-peroxo)dinickel(II) species and its eventual rapid decomposition to form the bis(μ-oxo)dinickel(III) complex. Also, the comparison of the formation of oxygen intermediates by nickel and copper complexes of the aforementioned ligands revealed that Ni ion facilitates homolysis of O-O bond of peroxo intermediate to form bis(μ-oxo)dinickel(III) species, conversely, Cu ion produces a stable (($\mu-\eta^2$:η^2-peroxo) complex. The [Ni^{III}(μ-O)$_2$ Ni^{III}] complexes of ligands (L6^X and L7-L9) are found to be stable at low temperature but slowly decomposes at elevated temperature and carry out the aliphatic hydroxylation reaction at the benzylic position of the ligand sidearm. However, the reactivity of complexes supported by dinucleating ligands (L7-L9) was lower than that of complexes stabilized by mononucleating ligands (L6^X) owing to the difficulty in the isomerization of [Ni^{II}(μ-OR)$_2$ Ni^{II}] complexes of former due to the presence of long alkyl straps. While probing the oxidation of external substrates such as 2,4-di-*tert*-butylphenol, 2,6-di-*tert*-butylphenol, 1,4-cyclohexadiene, triphenylphosphine and thioanisole it was found that [Ni^{III}(μ-O)$_2$Ni^{III}] complexes of L6^X and L7-L9 were efficiently abstracting the hydrogen atom from phenol and cyclohexadiene substrates and oxidizing them to corresponding diphenols and benzene respectively but did not oxidize triphenylphosphine and thioanisole. Moreover, these [Ni^{III}(μ-O)$_2$ Ni^{III}] complexes did not oxidize aromatic substrates as evident from the absence of ring hydroxylated product during the product analysis after the decomposition of bis(μ-oxo)dinickel(III) complex of dinucleating ligand L10 [76].

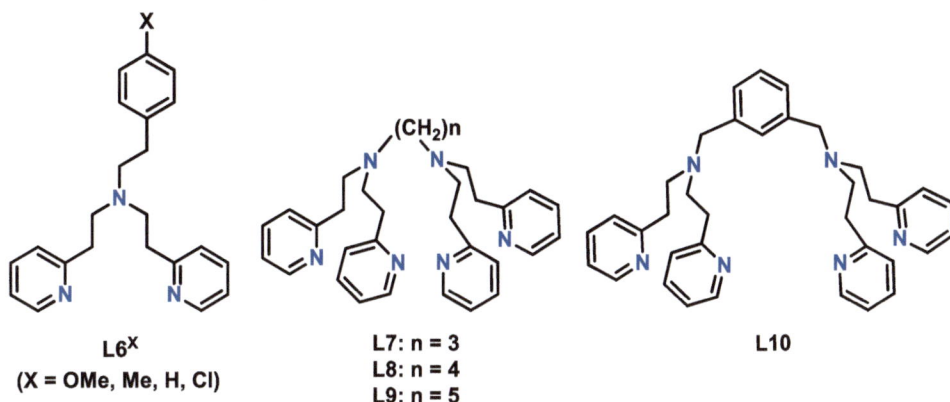

Fig. (3). The structures of mononucleating and dinucleating ligands (L6^x and L7-L10), Reprinted with permission from ref [76], Copyright 2001, American Chemical Society.

In 2000, Suzuki *et al.* demonstrated that the reaction of a bis(μ-hydroxo)dinickel(II) complex supported by the ligand Me$_3$-TPA (L11) (Fig. **4**) with H$_2$O$_2$ results in the formation of a bis(μ-oxo)dinickel(III) complex and that in turn reacts with an excess amount of H$_2$O$_2$ to afford a bis(μ-superoxo)dinickel(II) complex. The decomposition of these complexes in the presence of O$_2$ resulted in the oxidation of methyl groups of L11 to form carboxylate and alkoxide ligands [77]. In a subsequent paper, the author reported the formation of a bis(μ-alkylperoxo)dinickel(II) complex by the reaction of a bis(μ-hydroxo)dinickel(II) complex supported by Me$_2$-TPA ligand (L12) (Fig. **4**) with H$_2$O$_2$ and O$_2$ and this alkylperoxo species was found to be the plausible intermediate in the oxidation of methyl groups of L12 into carboxylate and alkoxide ligands. Single-crystal XRD study of the alkylperoxo species transpired that it has a Ni$_2$(μ-OOR)$_2$ core in which one of the methyl groups of each L12 ligand is oxidized to a ligand-based peroxide and the resulting two peroxides link to two nickel(II) ions (the chemical structure is given in Fig. (**5**) [78].

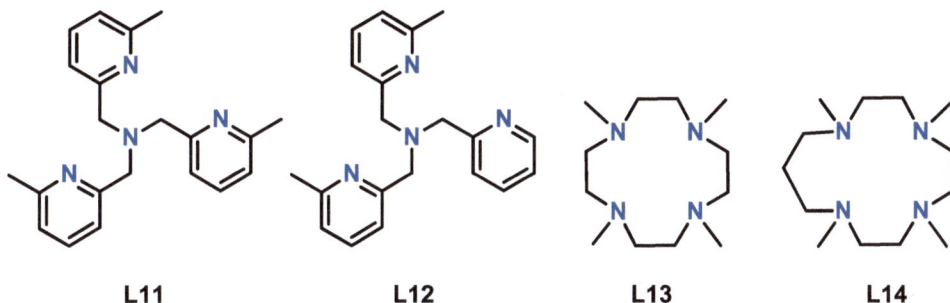

Fig. (4). The structures of ligands L11-L14 [77 - 80].

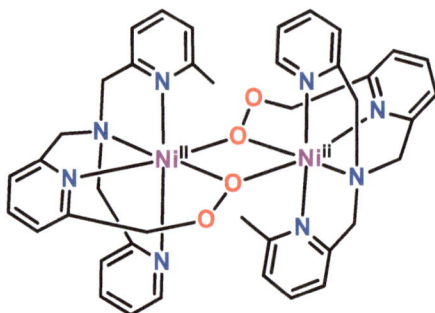

Fig. (5). The chemical structure of bis(μ-alkylperoxo)dinickel(II) complex [76].

After that, Solomon and Nam *et al.* characterized a side-on peroxo complex, **13b** Fig. (**6a**) having a Ni(III) center, prepared from the reaction of H_2O_2 with a Ni(II) precursor, **13a** stabilized by the 12-TMC ligand (L13) (Fig. **4**). Thermally stable **13b** adopted a distorted octahedral geometry and it was found to be inactive towards electrophilic reactions such as oxidation of PPh_3 and activated C-H substrates. However, **13b** was effective in carrying out nucleophilic reactions such as deformylation of aldehydes like 2-phenylpropionaldehyde (2-PPA) and cyclohexane carboxaldehyde (CCA). Moreover, a complete intermolecular transfer of O_2 was observed when $[Mn(II)(14\text{-}TMC)]^{2+}$ was added to **13b** as evidenced by the ESI-MS as well as spectroscopic analysis. The reaction in the presence of labeled ^{18}O confirmed that O_2 incorporated in $[Mn(III)(14\text{-}TMC)(O_2)]^+$ is not molecular oxygen [79]. It has been observed that the electronic, as well as the geometric structure of the $Ni\text{-}O_2$ complex, varies with the ring size of the TMC ligand. For instance, 14-TMC ligand facilitates the formation of a Ni(II)-superoxo complex whereas 12-TMC ligand favors the formation of Ni(III)-peroxo species [36]. However, a few years later, Nam *et al.* prepared mononuclear Ni(II)-superoxo (**14a**) and Ni(III)-peroxo (**14b**) complexes of 13-TMC ligand (L14) (Fig. **4**) by treating Ni(II) precursor complex, $[Ni^{II}(L14)(CH_3CN)]^{2+}$ with H_2O_2 in the presence of bases tetramethylammonium hydroxide (TMAH) and triethylamine (TEA), respectively. End-on coordination of O_2 in **14a** (Fig. **6b**) and side-on coordination of O_2 in **14b** (Fig. **6c**) was confirmed by both theoretical and spectroscopic studies. Similar to the previously reported superoxo complexes, **14a** was found to be active in electrophilic oxidation reactions such as oxygen atom transfer to PEt_3 to form $O=PEt_3$. On the other hand, **14b** was observed to be active in nucleophilic oxidation reactions such as deformylation of aldehyde like 2-phenylpropionaldehyde to form acetophenone [80].

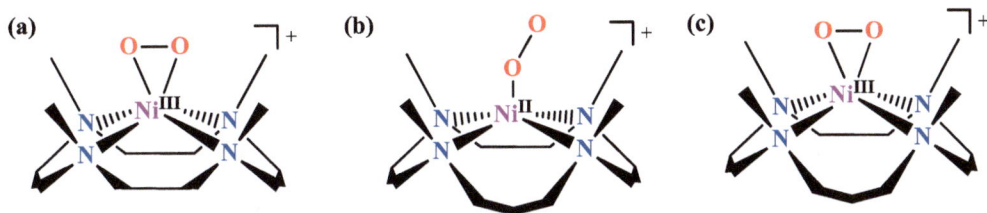

Fig. (6). The structures of (a) [Ni^{III}(L13)(O$_2$)]$^+$ (**13b**), (b) [Ni^{II}(L14)(O$_2$)]$^+$ (**14a**) and (c) [Ni^{III}(L14)(O$_2$)]$^+$ (**14b**) [79, 80].

In 2006, Itoh *et al.* reported the hydroxylation of alkanes using a Ni(II) complex, [Ni^{II}(L15)(OAc)(H$_2$O)]BPh$_4$ (**15**) of a tripodal ligand, L15 (Fig. 7) and *m*-CPBA as oxidant. High turnover numbers and excellent alcohol selectivity obtained after employing substrates such as cyclohexane, cyclooctane, adamantane, and ethylbenzene suggested the involvement of a NiO$^+$ (nickel-oxo) type catalytic intermediate. However, shreds of experimental or theoretical evidence to support the formation of suggested catalytic nickel-oxygen species were not provided. Complex **15** was found to be the most promising catalyst with respect to the turnover number as well as alcohol selectivity when compared with activities of Mn^{II}, Fe^{II}, and Co^{II} complexes of the same ligand, L15 [81]. In a subsequent paper from the same group, the author illustrated the ligand effect on catalytic activities of the complexes by employing Ni^{II} complexes of a series of pyridylalkylamine ligands, L16-L20 (Fig. 7) for cyclohexane hydroxylation in the presence of *m*-CPBA as oxidant. In order to study the effect of coordinated co-ligands, they synthesized a set of acetate complexes (**15a-20a**) and a set of nitrate complexes (**15b-18b**). Crystallographic studies revealed that the complexes except **15b** adopted distorted octahedral geometry with a mononuclear center whereas complex **15b** displayed a dimeric structure with distorted octahedral arrangement around the nickel center. Electronic absorption spectral behaviour ($^3A_{2g} \rightarrow ^3T_{2g}$(F) transition in the range 917-1010 nm, $^3A_{2g} \rightarrow ^3T_{1g}$(F) transition in the range 537-606 nm and, shoulder band of $^3A_{2g} \rightarrow ^1E_{1g}$(D) transition in the range 770-800 nm) of all the complexes except **20a** in CH$_2$Cl$_2$ confirmed the octahedral geometry of the complexes in solution as well. However, **20a** adopted a square pyramidal geometry with five weak d-d absorption bands ($^3B_1 \rightarrow ^3E$ = 389 nm, $^3B_1 \rightarrow ^3E$ = 634 nm, $^3B_1 \rightarrow ^3B_2$ = 796 nm, $^3B_1 \rightarrow ^3A_2$ = 831 nm, and $^3B_1 \rightarrow ^3E$ = 1350 nm) in a non-coordinating solvent like CH$_2$Cl$_2$. The pyridylmethylamine ligand complexes (**15a,17a**, and **19a**) were found to be more catalytically active than pyridylethylamine ligand complexes (**16a**, **18a**, and **20a**) and the enhanced catalytic activity of shorter alkyl chain bearing complexes is attributed to the more electron-donating ability of such systems. However, the alcohol selectivity obtained was much higher in the case of complexes with ethylene linker chain (**16a**, **18a**, and **20a**) than the complexes with methylene linker chain (**15a,17a**, and **19a)**. Electron donating ability of phenol-containing ligand systems was higher than ligands bearing pyridine donor moieties and hence the complexes of former ligand systems (**17a**, **17b**, **18a**, and **18b**) exhibited better catalytic efficiency. Denticity of the ligands also plays a crucial role in the catalytic activity of complexes. Complexes with tridentate ligands (**19a** and **20a**) were found to be less efficient than the complexes with tetradentate ligands (**15a** and **16a**). While investigating the effect of co-ligands on the catalytic efficiencies, the appearance of a lag phase was observed for the catalysis catalyzed by nitrate complexes. Although insight into the mechanistic features of the catalysis is not provided, Ni^{II}-$O^{\cdot}$ species generated *via* homolytic cleavage of *m*-CPBA adduct of Ni^{II} complex is proposed as the likely catalytic intermediate [82].

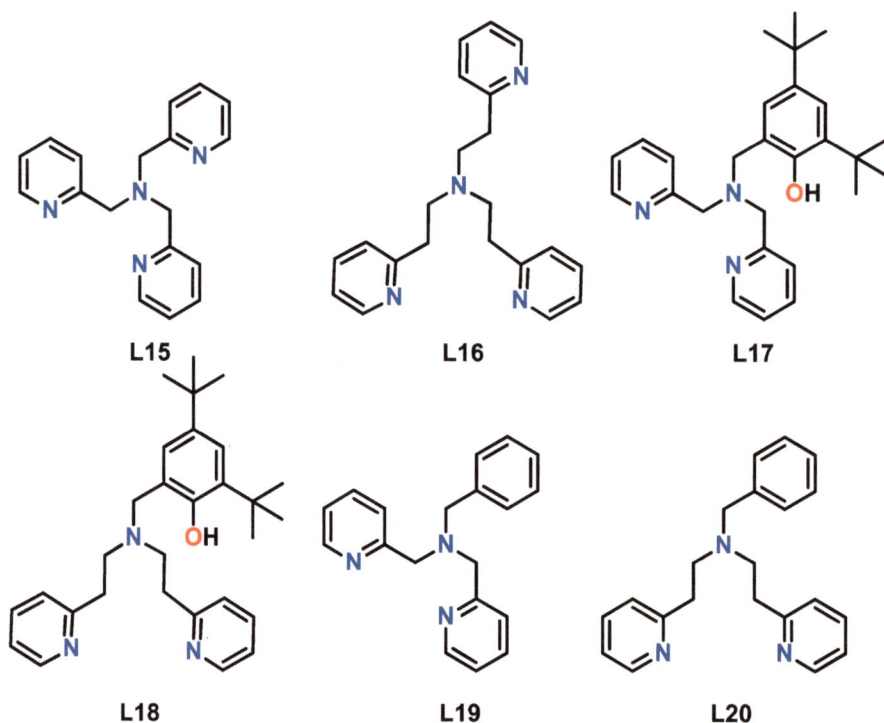

Fig. (7). The structures of ligands L15-L20 [81, 82].

In 2011, Palaniandavar and co-workers probed the influence of stereoelectronic effects of ligands on the catalytic efficiency of a series of Ni^{II} complexes of tetradentate tripodal ligands (L21-L27) (Figure **8**) in alkane oxidation reactions using *m*-CPBA as oxidant. X-ray crystallography revealed that complex [Ni(L22)(H$_2$O)(CH$_3$CN)](ClO$_4$)$_2$(**21a**) and complexes **22, 23,** and **24** having a general formula $[Ni(L)(CH_3CN)_2](BPh_4)_2$ (L22-L24) adopted a distorted octahedral environment around the nickel center in which each tripodal ligand is coordinated to the metal center through four nitrogen atoms and solvents such as CH_3CN or H_2O are occupied at the *cis* positions. Electronic spectra of all the Ni(II) complexes (**21a** and **21-27**) in a 3:1 solvent mixture of DCM:CH_3CN showcased the appearance of d-d bands in the visible region owing to the octahedral geometry of the complexes in the solution. Metal-ligand covalency and thus stereoelectronic properties of ligands, as well as Lewis acidity of the nickel center, play a pivotal role in the catalytic efficiency and product selectivity of oxidation of alkanes such as cyclohexane, adamantane, ethylbenzene, and cumene.

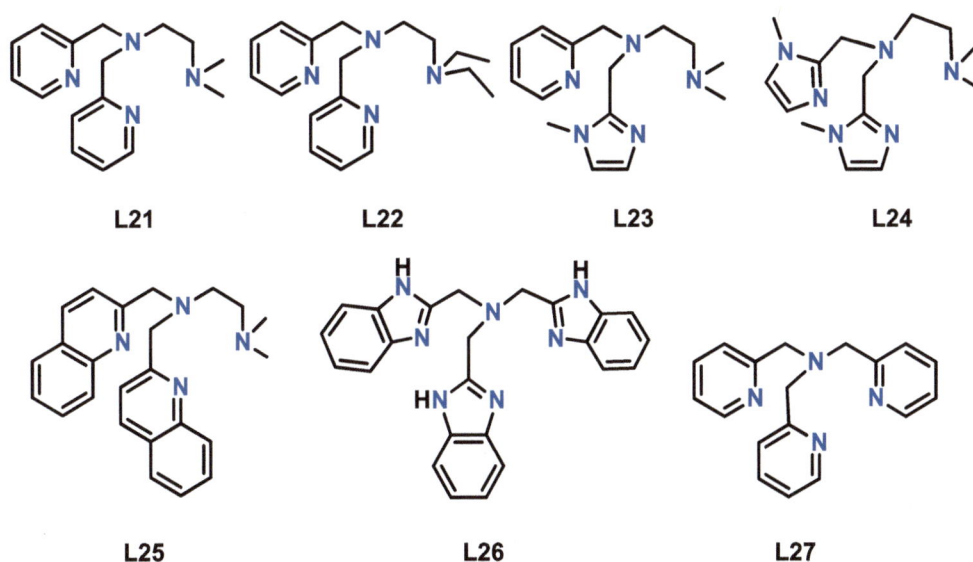

Fig. (8). The structures of tetradentate tripodal ligands (L21-L27) [83].

Variation in alcohol selectivity in accordance with the employment of catalysts **21-27** in oxidation reaction ascertained the involvement of metal-based oxygen intermediate rather than a free radical species. The plausible reactive intermediate $[(L)(CH_3CN)Ni-O^•]^+$ was assumed to be formed from the homolysis of O-O bond cleavage of oxidant bound adduct $[Ni^{II}(L)(CH_3CN)(OOCOC_6H_4Cl)]^+$. Complex **21** was found to be the most efficient catalyst among the family and catalytic efficiency was found to be decreasing with the introduction of bulkier substituents such as -NEt$_2$, quinolylmethyl group, and benzimidazolylmethyl group (L22, L25, and L26 respectively) in the ligand architecture. Also, the replacement of π-accepting pyridine donors by σ-bonding imidazole groups (L23-L24) to reduce the Lewis acidity of Ni(II) center resulted in a decrease in the catalytic performance of complexes (**23-24**) owing to the destabilization of nickel bound oxygen species. However, in the case of adamantane oxidation, a high selectivity ratio was observed while employing nickel complexes bearing sterically demanding substituents such as quinolylmethyl group or benzimidazolylmethyl group (L25-L26) [83]. Later, a combined experimental and theoretical study on alkane oxidation by employing a series of Ni(II) complexes supported by N5 ligands (L28-L32) (Fig. **9**) in the presence of *m*-CPBA [5].

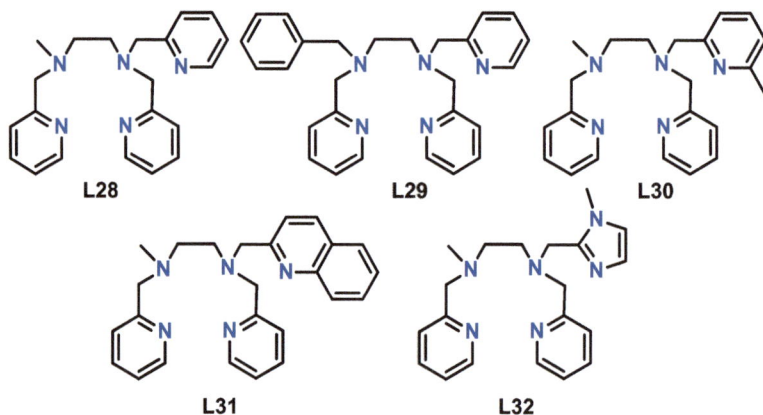

Fig. (9). The structures of ligands L28-L32. Reprinted with the permission from ref [5], Copyright 2014, Wiley-VCH Verlag GmbH & Co. KGaA, Weinheim.

All the complexes with an octahedral geometry in solution catalyzed the oxidation of substrates such as cyclohexane, adamantane, and cumene with appreciable selectivity and high TON. However, when compared to the catalytic activity of previously reported tetradentate N4 nickel complexes, nickel complexes of pentadentate N5 ligand series were found to be less efficient and this observation indicates that two coordination sites may be needed for better catalytic performance. The stereoelectronic effect was shown to affect the catalytic efficiencies as evident from the promising catalytic performance of complex stabilized by the ligand having π-back donating pyridine nitrogen atom (**28**) and the least performing complex bearing σ-donating imidazole nitrogen atom (**32**). DFT study of the catalytic mechanism has been demonstrated using cyclohexane and suggested that the oxidation occurs *via* a crucial pathway with the presence of a high-spin [(L)Ni^{II}-O$^{·}$]$^{+}$ species and two transition states, ts1$_{hs}$ and ts 2$_{doublet}$ [5].

Subsequently, they have reported another interesting study discussing the stereoelectronic effect of the ligand and solvent coordination on the catalytic potential of complexes. A family of Ni(II) complexes of tetradentate N4 ligands (L33-L37) (Fig. **10**) was isolated and studied for the alkane oxidation reaction. It is found that the complexes possessing diazacyclo ligand meridionally coordinate with Ni(II) and have only one labile axial site available for oxidant exchange, but catalyze cyclohexane and adamantane oxidation with high TON and selectivity. On the other hand, complex stabilized by ethylenediamine backbone (L37), which adopts *cis-α* or *cis-β* octahedral geometry depending on the solvent of coordination, supply two cis labile sites resulting in a better catalytic performance [10]. Hydroxylation of alkanes such as cyclohexane, adamantane, and cumene using a series of simple mixed ligand Ni(II) complexes and *m*-CPBA as an oxidant has also been demonstrated by the same author in which, picolinic acid (L38) is used as the primary ligand and bidentate *N,N'*-tetramethylethylenediamine (L39), 2,2'-bipyridine (L45), 1,10-phenanthroline (L47) and 2,9-dimethyl-1,10-phenanthroline (L48) and the tridentate *N,N',N''*pentamethyldiethylenetriamine (L40) as an ancillary ligand (Fig. **10** and Fig. **17**). Computational and electronic spectral studies revealed that all the complexes with a solvent (CH$_3$CN) coordination adopt a distorted octahedral geometry. However, octahedral geometry in the case of complex bearing L48 ligand is quite unstable and tends to adopt a stable square pyramidal geometry as evident from the more distorted *cis-* and *trans-* computational structures of this complex and a high β value of 20.7. Homolysis or heterolysis of O-O bond of *m*-CPBA-Ni(II) complex adduct to form high-valent nickel oxygen active species is proposed similar to the aforementioned reports. Also, the catalytic performance of complexes has an influence on denticity and steric encumbrance of the ligand. The complexes with tridentate L40 ligand or strongly π–back bonding planar L50 ligand exhibited an enhancement in the catalytic activity owing to the stability of the intermediate and a rapid ligand exchange process with oxidant as a result of a decrease in the Lewis acidity of Ni(II) center. On the other hand, among the complexes with bidentate co-ligands, the complexes with a non-planar L48 or L51 with sterically restraining methyl groups resulted in a decrease in catalytic performance [4].

In 2009, Suzuki and co-workers demonstrated the oxidation reactivity of a series of bis(μ- oxo)dinickel(III) complexes (**41-43**) of dinucleating ligands (L41-L43) (Fig. **11**) toward hydroxylation of aromatic ring strap in the ligand architecture. The generation of oxo complex having Ni$_2$O$_2$ core during the reaction of bis(μ-hydroxo)dinickel(II) complexes of same ligands and one equivalent of H$_2$O$_2$ was confirmed by the characteristic absorption band at 409 nm and vibrational rRaman spectral band at 616 cm^{-1} for Ni$_2$O$_2$ core. Product analysis after the decomposition of **41-43** revealed the formation of hydroxylated ligands which was confirmed by ^{1}H NMR spectral studies. The decomposition rate of bis(μ- oxo)dinickel(III) complexes was found to increase with an increase in the electron-releasing ability of the substituent attached to the aromatic ring of the supporting ligand, suggesting the involvement of intermediate during the oxidation reaction.

Moreover, the absence of a noticeable kinetic isotope effect (KIE ≈ 1) also ascertained that the electrophilic aromatic substitution pathway is more likely to occur rather than the hydrogen atom abstraction process [84].

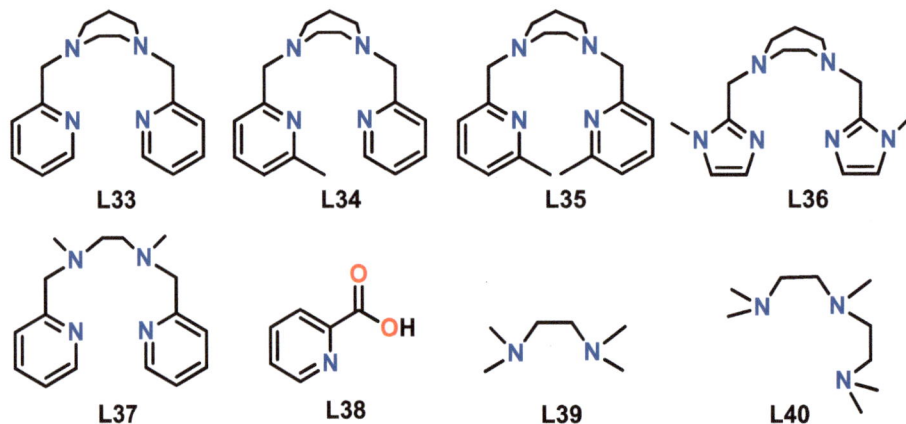

Fig. (10). The structures of ligands L33-L40. Reprinted with the permission from refs [4,10], Copyrights 2017, and 2013, Royal Society of Chemistry and Elsevier B.V.

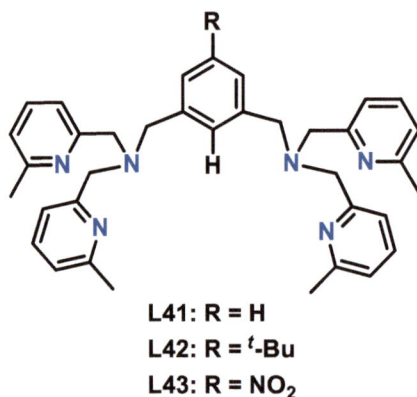

L41: R = H
L42: R = ^{t}-Bu
L43: R = NO$_2$

Fig. (11). The structures of ligands L41-L43 [84].

In 2015, McDonald and co-workers reported oxygen atom transfer and hydrogen atom abstraction reactivity of a thermally unstable, low-spin (S=1/2), square planar Ni^{III}-oxygen adduct, **44a** (Fig. **12**) derived from the 1:1 reaction of a Ni^{II}-bicarbonate complex (**44**) bearing a pyridinedicarboxamidate ligand (L44) with tris(4-bromophenyl)ammoniumyl hexachloroantimonate. The species **44a** was characterized by the appearance of two electronic spectral signatures at 520 and 720 nm irrespective of the solvent used, suggesting the absence of solvent coordination. An average g-value ($g_{av.}$=2.17) obtained from the X-band EPR analysis indicated that the unpaired electron is occupying the nickel center rather than in the ligand moiety and this observation of $g_\perp \gg g_\parallel$ suggested a square planar geometry for **44a**. XAS edge energy of 8345 eV obtained also supported the +3 oxidation of **44a**. Moreover, all the experimental findings were well supported by DFT calculations. Oxygen atom transfer reactivity of **44a** towards PPh$_3$ to form O=PPh$_3$ was detected by ESI-MS and ^{31}P NMR analysis. Also, oxidation of O-H bond of 2,6-di-*tert*-butyl phenol and the C-H bond of 1-benzyl-1,4-dihydronicotinamide (BNAH) was efficiently carried out by **44a** *via* hydrogen atom abstraction (HAA) as evidenced by a KIE value greater than 1 [85].

Fig. (12). The structure of **44a** [85].

In a subsequent paper, the author has illustrated the preparation of two other metastable high-valent nickel oxidants [Ni^{III}(OAc)(L44)] (**44b**) and [Ni^{III}(ONO$_2$)(L44)] (**44c**) bearing the same pyridinedicarboxamidate ligand (L44) by the addition of one-electron oxidant, magic blue to the corresponding Ni(II) square planar precursors and probed their reactivity in C-H bond activation. It has been observed that the oxidation of the OAc precursor complex could also be done using NaOCl/acetic acid and that of ONO$_2$ precursor could be done by the addition of cerium ammonium nitrate (CAN), which results in the formation of -OAc ligated **44b** and -ONO$_2$ ligated **44c**. A similar EPR and XAS spectral behavior to **44a** was observed for both **44b** and **44c** suggesting a similar geometry, as well as the occupancy of unpaired electron in the metal-based molecular orbital (d$_{xz}$ or d$_{z2}$). While probing the oxidative reactivity of these three complexes towards 2,6-di-*tert*-butylphenol, complex **44c** with electron-poor $^-$ONO$_2$ was shown to have a 15-fold higher activity than **44a** and **44b**. Because of the influence of electronic properties of ancillary ligands (-OAc, -CO$_2$H, -ONO$_2$) on the reactivity of the corresponding Ni^{III} complexes towards phenols, it is proposed that the association of the concerted proton-electron transfer pathway is also equally possible along with classical HAT mechanism (Fig. **13**). Among the three complexes, **44b** was found to be effective in oxidizing hydrocarbons with high C-H bond dissociation energy *via* a HAT mechanism as evident from a KIE value greater than 1 and a linear correlation between reaction rate and C-H bond dissociation energy [86].

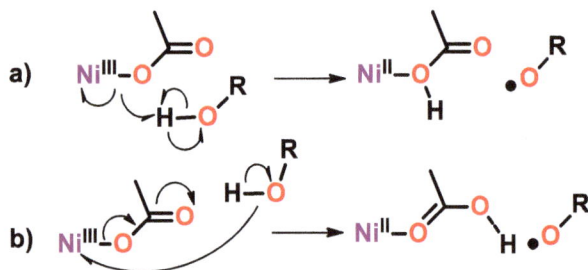

Fig. (13). (**a**) Classical HAT mechanism. (**b**) Concerted proton coupled and electron transfer. Reprinted with the permission from ref [86], Copyright 2016, American Chemical Society.

In 2017, the author illustrated the generation of a Ni^{III}-oxyl species by reaction of a Ni^{II} complex, $[Ni^{II}(NCCH_3)(L45)]$ (**45**) of a modified pyridinedicarboxamidate ligand, L45 (Fig. **14a**) and *m*-CPBA. Electronic absorption spectra with bands around 560 and 760 nm like the previously mentioned L44 supported complexes ascertained the formation of a Ni^{III} species and further studies using mass and EPR spectra (1:1 mixture of axial signal and rhombic signal) alluded to the formation of $[Ni^{III}(^-OOCC_6H_4Cl)(L45)]$ (**45a**). Similar to the preparation of **44a-c** by the one-electron oxidation process, **45a** could also be prepared by one-electron oxidation of $OOCC_6H_4Cl$ possessing Ni^{II} complex. GC-MS analysis of the reaction mixture of **45** and *m*-CPBA revealed the formation of *m*-CBA in high yield and thus indicated the two-electron oxidation of complex **45** by *m*-CPBA resulting from the heterolysis of O-O bond and **45a** is most likely to be derived from transient high valent nickel–oxygen species $Ni^{IV}=O/Ni^{III}-O^{\cdot}$ formed during the heterolytic O-O bond scission. This notion of O-O heterolysis is well supported by DFT calculations which also suggest that the Ni-oxygen species formed would be relatively stable $Ni^{III}-O^{\cdot}$ (S=1) rather than $Ni^{IV}=O$ (S=0). Organic product analysis after demetallation in the reaction mixture revealed the formation of a ligand oxidized product (Fig. **14b**) implying the promising oxidative power of this Ni^{III}-oxyl species towards a strong C-H bond. A crucial step in the formation of ligand oxidized product is the generation of methine radical resulting from the hydrogen atom abstraction of putative $Ni^{III}-O^{\cdot}$ from the methine position of the *iso*-propyl group [87]. Later, they synthesized a series of modified Ni^{II} complexes (**44d-i**) (Fig. **15**) bearing pyridinedicarboxamidate ligand as well as different ancillary ligands and performed one-electron oxidation of these complexes to form corresponding Ni^{III} complexes with S=1/2 state. While probing the reactivity of these Ni^{III} complexes towards phenol oxidation (2,6-di-*tert*-butylphenol), it is found that there exists a correlation between the oxidizing power of Ni^{III} oxidants and electron-donating ability of ancillary ligands.

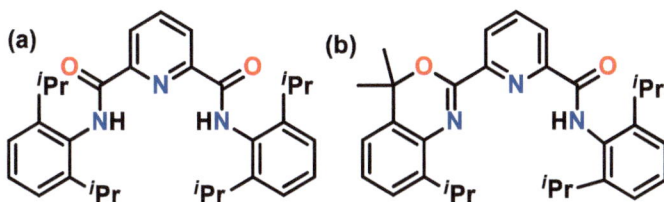

Fig. (14). The structure of ligand L45 (**a**) and ligand oxidized product (**b**) [87].

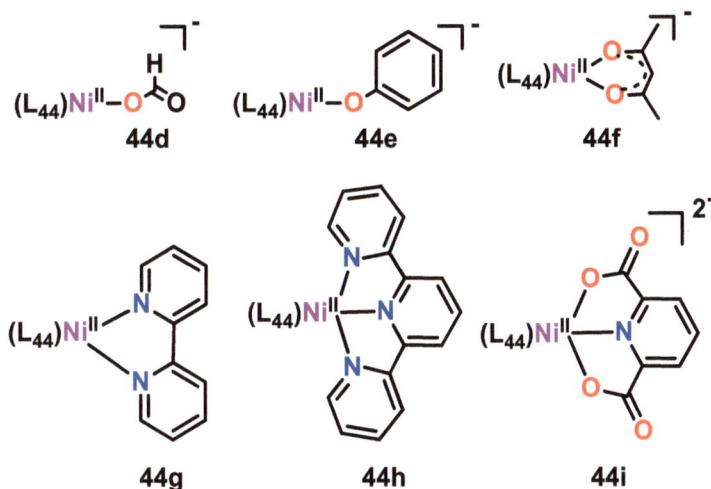

Fig. (15). The structure of Ni^{II} complexes of L44 with different ancillary ligands [88].

Complexes bearing highly basic supporting ligands were found to be inactive whereas those with neutral donors were active and thus the charge of the metal centre plays a crucial role in the reactivity [88]. Mechanistic investigation of this oxidation reaction was further executed using another Ni^{II} complex, [Ni^{II}(^{tBu}-terpy)(L44)] (**44j**) (^{tBu}-terpy =4,4′,4″-tri-*tert*-butyl-2,2′;6′,2″-terpyridine).

Reactivity of the corresponding Ni^{III} complex (**44j**) was probed using *para*-substituted 2,6-di-*tert*-butylphenol and GC-MS analysis of post-reaction mixture suggested proton-coupled electron transfer (PCET) reaction pathway owing to the presence of phenoxyl radical and proton accepted **44** (44-H$^+$) in the reaction mixture. However, unlike the previously mentioned Ni^{III} analogous, **44j** was found to be inactive towards hydrocarbons indicating its kinetic and thermodynamic inertness [88].

In 2017, Hartwig *et al.* demonstrated hydroxylation of polyethylenes (LDPE, HDPE, and LLDPE) by employing Ni catalyst of phenanthroline ligand (L52) and *m*-CPBA as oxidant. Among a series of nickel complexes, the selection of the most efficient complex [Ni(Me$_4$Phen)$_3$](BPh$_4$)$_2$ (**52**) was done by evaluating the catalytic potential of all the complexes of bi and tridentate nitrogen donor ligands (L46-L52) (Fig. **16**) in the oxidation of alkanes such as cyclohexane and *n*-octadecane. Complex **52** catalyzed the oxidation of cyclohexane with high alcohol to (ketone+ester) ratio of 10.5:1 and a TON of 5560 and the oxidation of *n*-octadecane with alcohol to ketone ratio of 15.5:1 and a TON of 660. The higher catalytic activity of complex **52** is attributed to the easy dissociation of phenanthroline ligand from the complex to coordinate *m*-CPBA. The application of this most efficient catalyst in polyethylene hydroxylation was successful with the incorporation of 2.0 to 5.5 functional groups (alcohol, ketone, alkyl chloride) per 100 monomer units.

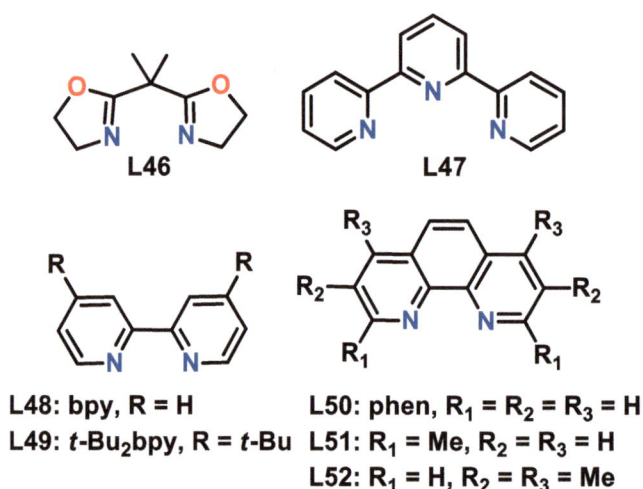

L46

L47

L48: bpy, R = H
L49: *t*-Bu$_2$bpy, R = *t*-Bu

L50: phen, R$_1$ = R$_2$ = R$_3$ = H
L51: R$_1$ = Me, R$_2$ = R$_3$ = H
L52: R$_1$ = H, R$_2$ = R$_3$ = Me

Fig. (16). The structure of ligands L48-L52. Adapted from ref [90].

The advantage of nickel catalyzed polyethylene hydroxylation over the conventional radical hydroxylation process is the absence of chain cleavage and cross-linking and thus a noticeable molecular weight change in the resulting polymer product is not observed [90]. Very recently, the same author has published an article on mechanistic investigation of nickel catalyzed oxidation of unactivated C(sp^3)-H bonds with *m*-CPBA. Contrary to several reports on the involvement of high-valent nickel oxygen species in the oxidation of C-H bonds, this mechanistic investigation points out that nickel complex only facilitates the decomposition of *m*-CPBA to form *m*-chlorobenzoyloxy radical, and thus the process of C-H bond oxidation of unactivated hydrocarbons is proceeding through the free radical mechanism. They employed a nickel complex of nitrogen-based ligand for the oxidation of cyclohexane and adamantane and it is observed that the stereoelectronic properties of ligands do not affect the yield or selectivity but as the steric constraints increase the time required for oxidation also increases. The formation of *m*-chlorobenzoyloxy radical responsible for C-H bond cleavage is indirectly identified by comparing the selectivity of this reaction with the selectivity of reactions mediated by other radical species such as hydroxyl, 3-chlorophenyl, and 3-chlorobenzoylperoxy. Hydrogen atom abstraction from the C-H bond of alkane by 3-chlorobenzoyloxy radical results in the formation of a carbon-centered long-lived radical as evidenced by CCl_4 trapping and radical clock (*cis*- and *trans*-1,2-di-methylcyclohexane) studies. Trapping of alkyl radical species by *m*-CPBA results in the formation of C-O bond and *m*-chlorobenzoyloxy species and thus continuing the propagation of radical chain [91].

Recently, Shanmugam *et al.* isolated a terminal trivalent Ni-OH complex, derived from the reaction of a nickel precursor, [Ni(COD)$_2$] (COD=cyclooctadiene) and a pincer ligand (L53) (Fig. **17**) in the presence of a trace amount of O_2. Multiple spectroscopic studies revealed the presence of terminal –OH group and paramagnetic nature and S = 1/2 ground state. Hydrogen atom abstraction of dihydroanthracene and dihydrobenzene by this high-valent oxidant also corroborated the presence of a proton bound to the terminal oxygen [92].

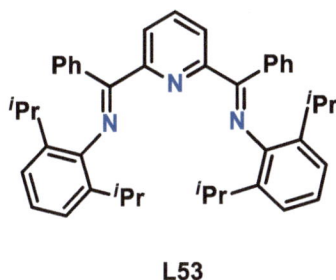

L53

Fig. (17). The structure of ligand L53, Reprinted with permission from ref [92], Copyright 2019, American Chemical Society.

Complexes Exhibiting Both Oxygen Atom Transfer and Hydrogen Atom Abstraction Reactivity

In 2012, Ray *et al.* reported the spectroscopic trapping of two metastable Ni(III) oxygen intermediates derived from the reaction of a Ni(II) complex, [Ni^{II}(TMG$_3$tren)]$^{2+}$ (**54**) of N4 tripodal ligand, TMG$_3$tren, (tris[2-(N-tetramethylguanidyl)ethyl] amine) (L54) with *m*-CPBA at low temperature. Electronic spectral absorption bands at 464 nm, 520 nm, and 794 nm, along with two rhombic S = 1/2 EPR signals with $g_\parallel > g_\perp$ values suggested the formation of high valent oxo species and generation of Ni(III) species, respectively. Ni(III)-oxo/-hydroxo intermediates are understood to be generated from the homolytic cleavage of O-O bond of terminal bound *m*-CPBA adduct of Ni(II) precursor complex (**54**). This high-valent nickel-oxygen intermediate was found to be reactive towards PPh$_3$ in oxygen atom transfer and efficiently abstracted the hydrogen atom from substrates such as 9,10-dihydroanthracene, 1,4-cyclohexadiene, xanthene and 1-benzyl-1,4-dihydronicotinamide [90]. Recently, a theoretical study about the structure, bonding, and reactivity of this putative Ni^{III} species has been reported and the oxyl radical character of the species has been revealed. Also, the radical nature of this species plays a major role in its robust reactivity towards hydrocarbon oxidation. Moreover, the possibility of involvement of two electromeric states, (i) Ni-O$^{\bullet}$ and (ii) Ni^{III}(O) differing mainly in Ni-O bond distance is also suggested [94].

In 2013, Hikichi *et al.* explored the influence of stereoelectronic properties of ligands on the catalytic activity of complexes by employing a series of Ni(II) complexes supported by tridentate substituted hydrotris(pyrazolyl)borate (Tp) ligands (L55-L58) (Fig. **18**) in cyclohexane oxidation with *m*-CPBA. A nickel(II)-acylperoxo complex ([Ni^{II}(OOC(=O)C$_6$H$_4$Cl)(Tp^{R2})]) derived from the stoichiometric reaction of nickel(II)-bis(μ-hydroxo) complexes of L55-L58 with *m*-CPBA was assumed to be the active oxidant species in the catalysis by analyzing the UV-Vis spectra with an absorption band around 386 nm and an IR spectral peak at 1644 cm^{-1}. These putative acylperoxo complexes were found to be thermally unstable as evident from the disappearance of the UV-Vis band and IR peak at a higher temperature. The incorporation of an electron-withdrawing group like bromine on the pyrazolyl backbone of ligand (L58) resulted in an enhancement in the electrophilicity of active oxidant as well as an increment in alcohol selectivity.

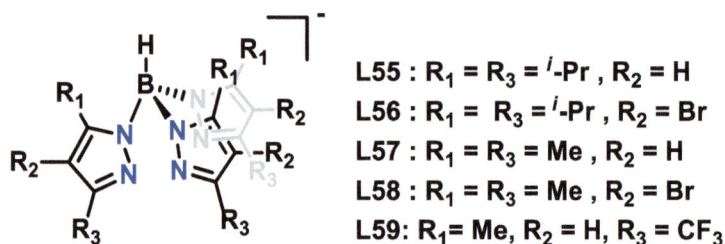

L55 : R$_1$ = R$_3$ = ^{i}-Pr , R$_2$ = H
L56 : R$_1$ = R$_3$ = ^{i}-Pr , R$_2$ = Br
L57 : R$_1$ = R$_3$ = Me , R$_2$ = H
L58 : R$_1$ = R$_3$ = Me , R$_2$ = Br
L59 : R$_1$ = Me, R$_2$ = H, R$_3$ = CF$_3$

Fig. (18). The structure of ligands L55-L59, Reprinted with the permission from ref [97], Copyright 2013, American Chemical Society.

Ni(II) complex bearing most sterically encumbered ligands L55 and L56 exhibited almost no catalytic activity towards cyclohexane oxygenation whereas less sterically demanding L57 and L58 bearing complexes displayed higher catalytic activity to yield hydroxylated cyclohexane. However, a ^{1}H NMR signal at 16.5 ppm of the putative acylperoxo complexes of L55 and L56 revealed the intramolecular ligand hydroxylation at the methane portion of the isopropyl group of L55 and L56 [95]. An alkylperoxo complex, $[Ni^{II}(OOtBu)(Tp^{iPr})]$ **55b** derived from the reaction of $[(Ni^{II}Tp^{iPr})_2(\mu\text{-OH})_2]$ with a stoichiometric amount of *tert*-butylhydroperoxide was found to exhibit either electrophilicity (with PPh_3 or CO) or nucleophilicity (with ArCHO) depending on the nature of substrates and oxidizes alkanes such as cyclohexane presumably *via* a radical mechanism to afford alcohol and ketone with low alcohol to ketone ratio. The generation of radical species through O-O and Ni-O bond homolysis was proposed in the mechanism [96]. In a subsequent paper, the author modified one of the alkyl substituents on the pyrazolyl backbone to an electron-withdrawing, oxidation resistant, and moderately sterically restrained trifluoromethyl groups (L59) (Figure **18**) and successfully isolated corresponding acylperoxo Ni(II) complex **59**. Complex **59** was found to be in a distorted square pyramidal geometry ($\tau =$ 0.26) with a high spin (S = 1) Ni(II) center and unlike the previously reported acylperoxo Ni(II) complexes of Tp ligands (L55-L58) displayed thermal stability even at 70° C owing to the steric hindrance around the Ni center as well as the electron-withdrawing nature of CF_3. To probe the oxidizing reactivity of complex **59** in detail, external substrates such as sulfides, phosphines, alkenes, and hydrocarbons with activated C-H bonds were employed. The complex **59** with an electrophilic nature was found to be efficient in transferring an oxygen atom to various *p*-substituted thioanisoles, phosphines, and alkenes such as cyclohexene and styrene. The absence of absorption band around 800 nm in the UV-Vis spectrum obtained during the oxidation of activated C-H bond having hydrocarbons like 1,4-cyclohexadiene, 9,10-dihydroanthracene, xanthene, and fluorene suggested the absence of O-O homolysis in complex **59** to from high-valent Ni-oxygen species. On the other hand, oxidation of unactivated C-H substrates like cyclohexane was expected to occur mainly *via* the formation of high-valent oxygen species by homolytic cleavage of O-O bond in the acylperoxo nickel complexes of L55-L58 [97].

CONCLUSION

In this book chapter, we have summarized noteworthy attempts in oxo atom transfer reactions catalyzed by various nickel complexes in the presence of oxidants such as *m*-CPBA, H_2O_2, *t*-BuOOH, and molecular oxygen. Interestingly, most of the complexes exhibited promising reactivity in either electrophilic oxidation (hydrogen atom abstraction, oxygen atom transfer) and/or nucleophilic oxidation reactions. The importance of ligand design is evident from the observed reactivity of complexes as reactivity is correlated to stereoelectronic properties of ligand architectures. In the case of dinuclear nickel complexes, bis(μ-oxo)dinickel(III), bis(μ- peroxo)dinickel(II), and bis(μ- superoxo)dinickel(II) complexes are proposed to be the plausible intermediate and in mononuclear complexes putative high-valent nickel-oxo/-oxyl species generated *via* heterolysis or homolysis of O-O bond of oxidant adduct of precursor nickel complex is proposed to be the likely intermediate. However, elucidation of the actual species and providing a clear-cut mechanistic description still remains elusive. Also, the recent finding states that oxidation of unactivated $C(sp^3)$-H bonds do not involve metal-based oxo/oxyl species, which is contrary to the previously reported oxidation mechanisms and thus draws a great deal of attention as well as urge the scientific community to bring forth the strategies to isolate or completely characterize the putative intermediate species in such reactions.

CONSENT FOR PUBLICATION

Not applicable.

CONFLICT OF INTEREST

The author declares no conflict of interest, financial or otherwise.

ACKNOWLEDGEMENTS

Sankaralingam, M. sincerely acknowledges the department of science and technology for the award of the DST-Inspire Faculty research grant (IFA-17-CH286) and also the National Institute of Technology Calicut for the Faculty Research Grant. AR acknowledges the UGC-JRF for the fellowship.

REFERENCES

[1] Sankaralingam, M.; Balamurugan, M.; Palaniandavar, M. Alkane and alkene oxidation reactions catalyzed by nickel(II) complexes: Effect of ligand factors. *Coord. Chem. Rev.,* **2020**, *403*, 213085.
[http://dx.doi.org/10.1016/j.ccr.2019.213085]

[2] Mayilmurugan, R.; Sankaralingam, M.; Suresh, E.; Palaniandavar, M. Novel square pyramidal iron(III) complexes of linear tetradentate bis(phenolate) ligands as structural and reactive models for intradiol-cleaving 3,4-PCD enzymes: Quinone formation *vs.* intradiol cleavage. *Dalton Trans.,* **2010**, *39*(40), 9611-9625.
[http://dx.doi.org/10.1039/c0dt00171f] [PMID: 20835480]

[3] Sundaravel, K.; Sankaralingam, M.; Suresh, E.; Palaniandavar, M. Biomimetic iron(III) complexes of N_3O and N_3O_2 donor ligands: protonation of coordinated ethanolate donor enhances dioxygenase activity. *Dalton Trans.,* **2011**, *40*(33), 8444-8458.
[http://dx.doi.org/10.1039/c1dt10495k] [PMID: 21785763]

[4] Sankaralingam, M.; Vadivelu, P.; Suresh, E. Palaniandavar. M. Mixed ligand nickel(II) complexes as catalysts for alkane hydroxylation using m-chloroperbenzoic acid as oxidantInorg. *Chim. Acta.,* **2013**, *407*, 98-107.

[5] Sankaralingam, M.; Balamurugan, M.; Palaniandavar, M.; Vadivelu, P.; Suresh, C.H. Nickel(II) complexes of pentadentate N5 ligands as catalysts for alkane hydroxylation by using m-CPBA as oxidant: a combined experimental and computational study. *Chemistry,* **2014**, *20*(36), 11346-11361.
[http://dx.doi.org/10.1002/chem.201402391] [PMID: 25100547]

[6]   Sankaralingam, M.; Palaniandavar, M. Tuning the olefin epoxidation by manganese(III) complexes of bisphenolate ligands: effect of Lewis basicity of ligands on reactivity. *Dalton Trans.,* **2014**, *43*(2), 538-550.
[http://dx.doi.org/10.1039/C3DT51766G] [PMID: 24121481]

[7]   Sankaralingam, M.; Palaniandavar, M. Diiron(III) complexes of tridentate 3N ligands as functional models for methane monooxygenases: Effect of the capping ligand on hydroxylation of alkanes. *Polyhedron,* **2014**, 171-180.
[http://dx.doi.org/10.1016/j.poly.2013.08.067]

[8]   Saravanan, N. Sankaralingama. M.; Palaniandavar, M. Manganese(II) complexes of tetradentate 4N ligands with diazepane backbones for catalytic olefin epoxidation: effect of nucleophilicity of peroxo complexes on reactivity. *RSC Advances,* **2014**, *4*, 12000-12011.
[http://dx.doi.org/10.1039/C3RA44729D]

[9]   Sankaralingam, M.; Jeon, S.H.; Lee, Y.M.; Seo, M.S.; Ohkubo, K.; Fukuzumi, S.; Nam, W. An amphoteric reactivity of a mixed-valent bis(μ-oxo)dimanganese(III,IV) complex acting as an electrophile and a nucleophile. *Dalton Trans.,* **2016**, *45*(1), 376-383.
[http://dx.doi.org/10.1039/C5DT04292E] [PMID: 26620273]

[10]   Sankaralingam, M.; Vadivelu, P.; Palaniandavar, M. Novel nickel(ii) complexes of sterically modified linear N4 ligands: effect of ligand stereoelectronic factors and solvent of coordination on nickel(ii) spin-state and catalytic alkane hydroxylation. *Dalton Trans.,* **2017**, *46*(22), 7181-7193.
[http://dx.doi.org/10.1039/C7DT00576H] [PMID: 28418046]

[11]   Sankaralingam, M.; Lee, Y.; Nam, W.; Fukuzumi, S. Amphoteric reactivity of metal–oxygen complexes in oxidation Reactions. *Coord. Chem. Rev.,* **2018**, *365*, 41-59.
[http://dx.doi.org/10.1016/j.ccr.2018.03.003]

[12]   Hong, S.; Lee, Y.M.; Sankaralingam, M.; Vardhaman, A.K.; Park, Y.J.; Cho, K.B.; Ogura, T.; Sarangi, R.; Fukuzumi, S.; Nam, W.; Manganese, A. (V)–oxo complex: Synthesis by dioxygen activation and enhancement of its oxidizing power by binding scandium ion. *J. Am. Chem. Soc.,* **2016**, *138*(27), 8523-8532.
[http://dx.doi.org/10.1021/jacs.6b03874] [PMID: 27310336]

[13]   Devi, T.; Lee, Y.M.; Jung, J.; Sankaralingam, M.; Nam, W.; Fukuzumi, S. Angew. A Chromium(III)-superoxo complex as a three-electron oxidant with a large tunneling effect in multi-electron oxidation of NADH analogues. *Angew. Chem. Int. Ed. Engl.,* **2017**, *56*(13), 3510-3515.
[http://dx.doi.org/10.1002/anie.201611709] [PMID: 28266771]

[14]   Sankaralingam, M.; Lee, Y.M.; Nam, W.; Fukuzumi, S. Selective oxygenation of cyclohexene by dioxygen *via* an iron(V)-oxo complex-autocatalyzed reaction. *Inorg. Chem.,* **2017**, *56*(9), 5096-5104.
[http://dx.doi.org/10.1021/acs.inorgchem.7b00220] [PMID: 28422498]

[15]   Sankaralingam, M.; Lee, Y.M.; Lu, X.; Vardhaman, A.K.; Nam, W.; Fukuzumi, S. Autocatalytic dioxygen activation to produce an iron(V)-oxo complex without any reductants. *Chem. Commun. (Camb.),* **2017**, *53*(59), 8348-8351.
[http://dx.doi.org/10.1039/C7CC03742B] [PMID: 28696437]

[16]   Sankaralingam, M.; Lee, Y.; Jeon, S.H.; Seo, M.S.; Cho, K. Nam. W. A manganese(V)–oxo tetraamido macrocyclic ligand (TAML) cation radical complex: Synthesis, characterization, and reactivity studies. *Chem. Commun. (Camb.),* **2018**, *54*, 1209-1212.
[http://dx.doi.org/10.1039/C7CC09492B] [PMID: 29335701]

[17]   Karmalkar, D.G.; Li, X.X.; Seo, M.S.; Sankaralingam, M.; Ohta, T.; Sarangi, R.; Hong, S.; Nam, W. Nam. W. A manganese(V)–oxo tetraamido macrocyclic ligand (TAML) cation radical complex: Synthesis, characterization, and reactivity studies. *Chemistry,* **2018**, *24*(68), 17927-17931.
[http://dx.doi.org/10.1002/chem.201804898] [PMID: 30267428]

[18]   Saracini, C.; Malik, D.D.; Sankaralingam, M.; Lee, Y.M.; Nam, W.; Fukuzumi, S. Enhanced electron-transfer reactivity of a long-lived photoexcited state of a cobalt–oxygen complex. *Inorg. Chem.,* **2018**,

57(17), 10945-10952.
[http://dx.doi.org/10.1021/acs.inorgchem.8b01571] [PMID: 30133298]

[19]   Sankaralingam, M.; Lee, Y.M.; Karmalkar, D.G.; Nam, W.; Fukuzumi, S. A Mononuclear non-heme manganese(III)–aqua complex as a new active oxidant in hydrogen atom transfer reactions. *J. Am. Chem. Soc.,* **2018**, *140*(40), 12695-12699.
[http://dx.doi.org/10.1021/jacs.8b07772] [PMID: 30269497]

[20]   Sankaralingam, M.; Lee, Y.M.; Pineda-Galvan, Y.; Karmalkar, D.G.; Seo, M.S.; Jeon, S.H.; Pushkar, Y.; Fukuzumi, S.; Nam, W. Redox reactivity of a manganese-oxo complex binding calcium ion and other redox-inactive metal ions. *J. Am. Chem. Soc.,* **2019**, *141*(3), 1324-1336.
[http://dx.doi.org/10.1021/jacs.8b11492] [PMID: 30580510]

[21]   Harmalkar, D.S.; Santosh, G.; Shetgaonkar, S.B.; Sankaralingam, M.; Dhuri, S.N. A putative heme manganese(V)- oxo species in the C–H activation and epoxidation reactions in an aqueous buffer. *New J. Chem.,* **2019**, *43*, 12900-12906.
[http://dx.doi.org/10.1039/C9NJ01381D]

[22]   Lu, X.; Lee, Y.; Sankaralingam, M.; Fukuzumi, S. Nam. W. Selective Oxygenation of Cyclohexene by Dioxygen *via* an Iron(V)-Oxo Complex-Autocatalyzed Reaction. *Inorg. Chem.,* **2020**, *59*, 18010-18017.
[http://dx.doi.org/10.1021/acs.inorgchem.0c02400] [PMID: 33300784]

[23]   Narulkar, D.D.; Ansari, A.; Vardhaman, A.K.; Harmalkar, S.S.; Lingamallu, G.; Dhavale, V.M.; Sankaralingam, M.; Das, S.; Kumar, P.; Dhuri, S.N. A side-on Mn(III)-peroxo supported by a non-heme pentadentate N_3Py_2 ligand: synthesis, characterization and reactivity studies. *Dalton Trans.,* **2021**, *50*(8), 2824-2831.
[http://dx.doi.org/10.1039/D0DT03706K] [PMID: 33533342]

[24]   Das, A.; Rajeev, A.; Bhunia, S.; Arunkumar, M.; Chari, N.; Sankaralingam, M. Synthesis, characterization and antimicrobial activity of nickel(II) complexes of tridentate N3 ligands. *Inorg. Chim. Acta,* **2021**, *526*, 120515.
[http://dx.doi.org/10.1016/j.ica.2021.120515]

[25]   Mohammed, T.P.; Sankaralingam, M. Reactivities of high valent manganese-oxo porphyrins in aqueous medium. *Tetrahedron,* **2022**, *103*, 132483.
[http://dx.doi.org/10.1016/j.tet.2021.132483]

[26]   Nam, W. High-valent iron(IV)-oxo complexes of heme and non-heme ligands in oxygenation reactions. *Acc. Chem. Res.,* **2007**, *40*(7), 522-531.
[http://dx.doi.org/10.1021/ar700027f] [PMID: 17469792]

[27]   Barona-Castaño, J.C.; Carmona-Vargas, C.C.; Brocksom, T.J.; de Oliveira, K.T. Porphyrins as catalysts in scalable organic reactions. *Molecules,* **2016**, *21*(3), 310.
[http://dx.doi.org/10.3390/molecules21030310] [PMID: 27005601]

[28]   Groves, J.T.; Nemo, T.E.; Myers, R.S. Hydroxylation and epoxidation catalyzed by iron-porphine complexes. Oxygen transfer from iodosylbenzene. *J. Am. Chem. Soc.,* **1979**, *101*, 1032-1033.
[http://dx.doi.org/10.1021/ja00498a040]

[29]   Meunier, B. Metalloporphyrins as versatile catalysts for oxidation reactions and oxidative DNA cleavage. *Chem. Rev.,* **1992**, *92*, 1411-1456.
[http://dx.doi.org/10.1021/cr00014a008]

[30]   Rahimi, R.; Gholamrezapor, E.; Naimi-jamal, M.R. Oxidation of benzyl alcohols to the corresponding carbonyl compounds catalyzed by copper (II) meso-tetra phenyl porphyrin as cytochrome P-450 model reaction. *Inorg. Chem. Commun.,* **2011**, *14*, 1561-1568.
[http://dx.doi.org/10.1016/j.inoche.2011.05.056]

[31]   Nam, W.; Han, H.J.; Oh, S.; Lee, Y.J.; Choi, M.; Han, S.; Kim, C.; Woo, S.K.; Shin, W.J. New insights into the mechanisms of O–O bond cleavage of hydrogen peroxide and *tert*-alkyl hydroperoxides by iron(III) porphyrin complexes. *Am. Chem. Soc.,* **2000**, *122*, 8677-8684.

[http://dx.doi.org/10.1021/ja994403e]

[32]   Zhang, L.; Lee, Y.; Guo, M.; Fukuzumi, S.; Nam, W. Enhanced redox reactivity of a nonheme iron(V)–oxo complex binding proton. *J. Am. Chem. Soc.,* **2020,** *142,* 19879-19884.
[http://dx.doi.org/10.1021/jacs.0c10159] [PMID: 33186008]

[33]   Ragsdale, S.W. Nickel-based enzyme systems. *J. Biol. Chem.,* **2009,** *284*(28), 18571-18575.
[http://dx.doi.org/10.1074/jbc.R900020200] [PMID: 19363030]

[34]   Wang, W.J.; Wei, W.J.; Liao, R.Z. Deciphering the chemoselectivity of nickel-dependent quercetin 2,4-dioxygenase. *Phys. Chem. Chem. Phys.,* **2018,** *20*(23), 15784-15794.
[http://dx.doi.org/10.1039/C8CP02683A] [PMID: 29869653]

[35]   de Visser, S.P.; Rohde, J.; Lee, Y.; Cho, J.; Nam, W. Intrinsic properties and reactivities of mononuclear nonheme iron–oxygen complexes bearing the tetramethylcyclam ligand. *Coord. Chem. Rev.,* **2013,** *257,* 381-393.
[http://dx.doi.org/10.1016/j.ccr.2012.06.002]

[36]   Cho, J.; Sarangi, R.; Nam, W. Mononuclear metal-O$_2$ complexes bearing macrocyclic *N*-tetramethylated cyclam ligands. *Acc. Chem. Res.,* **2012,** *45*(8), 1321-1330.
[http://dx.doi.org/10.1021/ar3000019] [PMID: 22612523]

[37]   Nam, W.; Lee, Y.M.; Fukuzumi, S. Tuning reactivity and mechanism in oxidation reactions by mononuclear nonheme iron(IV)-oxo complexes. *Acc. Chem. Res.,* **2014,** *47*(4), 1146-1154.
[http://dx.doi.org/10.1021/ar400258p] [PMID: 24524675]

[38]   Nam, W. Synthetic mononuclear nonheme iron-oxygen intermediates. *Acc. Chem. Res.,* **2015,** *48*(8), 2415-2423.
[http://dx.doi.org/10.1021/acs.accounts.5b00218] [PMID: 26203519]

[39]   Ray, K.; Heims, F.; Schwalbe, M.; Nam, W. High-valent metal-oxo intermediates in energy demanding processes: from dioxygen reduction to water splitting. *Curr. Opin. Chem. Biol.,* **2015,** *25,* 159-171.
[http://dx.doi.org/10.1016/j.cbpa.2015.01.014] [PMID: 25703840]

[40]   Ray, K.; Pfaff, F.F.; Wang, B.; Nam, W. Status of reactive non-heme metal-oxygen intermediates in chemical and enzymatic reactions. *J. Am. Chem. Soc.,* **2014,** *136*(40), 13942-13958.
[http://dx.doi.org/10.1021/ja507807v] [PMID: 25215462]

[41]   Cho, K.B.; Hirao, H.; Shaik, S.; Nam, W. To rebound or dissociate? This is the mechanistic question in C-H hydroxylation by heme and nonheme metal-oxo complexes. *Chem. Soc. Rev.,* **2016,** *45*(5), 1197-1210.
[http://dx.doi.org/10.1039/C5CS00566C] [PMID: 26690848]

[42]   Hong, S.; Lee, Y.; Ray, K.; Nam, W. Making and breaking of the O-O bond at iron complexes. *Coord. Chem. Rev.,* **2017,** *334,* 25-42.
[http://dx.doi.org/10.1016/j.ccr.2016.07.006]

[43]   Nam, W.; Lee, Y.M.; Fukuzumi, S. Hydrogen atom transfer reactions of mononuclear nonheme metal–oxygen intermediates. *Acc. Chem. Res.,* **2018,** *51*(9), 2014-2022.
[http://dx.doi.org/10.1021/acs.accounts.8b00299] [PMID: 30179459]

[44]   Guo, M.; Corona, T.; Ray, K.; Nam, W. Heme and nonheme high-valent iron and manganese oxo cores in biological and abiological oxidation reactions. *ACS Cent. Sci.,* **2019,** *5*(1), 13-28.
[http://dx.doi.org/10.1021/acscentsci.8b00698] [PMID: 30693322]

[45]   Larson, V.A.; Battistella, B.; Ray, K.; Lehnert, N.; Nam, W. Nat. Iron and manganese oxo complexes, oxo wall and beyond. *Rev. Chem.,* **2020,** *4,* 404-419.

[46]   Costas, M.; Mehn, M.P.; Jensen, M.P.; Que, L., Jr Dioxygen activation at mononuclear nonheme iron active sites: enzymes, models, and intermediates. *Chem. Rev.,* **2004,** *104*(2), 939-986.
[http://dx.doi.org/10.1021/cr020628n] [PMID: 14871146]

[47] Que, L., Jr The road to non-heme oxoferryls and beyond. *Acc. Chem. Res.,* **2007**, *40*(7), 493-500.
[http://dx.doi.org/10.1021/ar700024g] [PMID: 17595051]

[48] McDonald, A.R.; Que, L., Jr High-valent nonheme iron-oxo complexes: Synthesis, structure, and spectroscopy. *Coord. Chem. Rev.,* **2013**, *257*, 414-428.
[http://dx.doi.org/10.1016/j.ccr.2012.08.002]

[49] Oloo, W.N.; Que, L., Jr Bioinspired nonheme iron catalysts for C–H and C=C bond oxidation: Insights into the nature of the metal-based oxidan. *Acc. Chem. Res.,* **2015**, *48*(9), 2612-2621.
[http://dx.doi.org/10.1021/acs.accounts.5b00053] [PMID: 26280131]

[50] Puri, M.; Que, L., Jr Toward the synthesis of more reactive *S* = 2 non-heme oxoiron(IV) complexes. *Acc. Chem. Res.,* **2015**, *48*(8), 2443-2452.
[http://dx.doi.org/10.1021/acs.accounts.5b00244] [PMID: 26176555]

[51] Que, L., Jr; Puri, M. The Amazing High-valent nonheme iron-oxo landscape. *Bull. Jpn. Soc. Coord. Chem.,* **2016**, *67*, 10-18.
[http://dx.doi.org/10.4019/bjscc.67.10]

[52] Kal, S.; Que, L. Dioxygen activation by nonheme iron enzymes with the 2-His-1-carboxylate facial triad that generate high-valent oxoiron oxidants. *Eur. J. Biochem.,* **2017**, *22*(2-3), 339-365.
[http://dx.doi.org/10.1007/s00775-016-1431-2] [PMID: 28074299]

[53] Raven, E.L. Oxygen activation by mononuclear Mn, Co, and Ni centers in biology and synthetic complexes. *J. Biol. Inorg. Chem.,* **2017**, *22*, 175-183.
[http://dx.doi.org/10.1007/s00775-016-1412-5] [PMID: 27909919]

[54] Huang, X.; Groves, J.T. Beyond ferryl-mediated hydroxylation: 40 years of the rebound mechanism and C-H activation. *Eur. J. Biochem.,* **2017**, *22*(2-3), 185-207.
[http://dx.doi.org/10.1007/s00775-016-1414-3] [PMID: 27909920]

[55] Que, L., Jr 60 years of dioxygen activation. *Eur. J. Biochem.,* **2017**, *22*(2-3), 171-173.
[http://dx.doi.org/10.1007/s00775-017-1443-6] [PMID: 28190114]

[56] Jasniewski, A.J.; Que, L., Jr Dioxygen activation by nonheme diiron enzymes: Diverse dioxygen adducts, high- valent intermediates, and related model complexes. *Chem. Rev.,* **2018**, *118*(5), 2554-2592.
[http://dx.doi.org/10.1021/acs.chemrev.7b00457] [PMID: 29400961]

[57] Klein, J.E.M.N.; Que, L., Jr *Biomimetic high-valent mononuclear nonheme iron-oxo chemistry. Encyclopedia of Inorganic and Bioinorganic Chemistry*; John Wiley and Sons, Inc.: Hoboken, **2016**.

[58] Kal, S.; Xu, S.; Que, L., Jr Bio-inspired nonheme iron oxidation catalysis: Involvement of oxoiron(V) oxidants in cleaving strong C–H bonds. *Angew. Chem. Int. Ed. Engl.,* **2020**, *59*(19), 7332-7349.
[http://dx.doi.org/10.1002/anie.201906551] [PMID: 31373120]

[59] Gamez, P.; Aubel, P.G.; Driessen, W.L.; Reedijk, J. Homogeneous bio-inspired copper-catalyzed oxidation reactions. *Chem. Soc. Rev.,* **2001**, *30*, 376-385.
[http://dx.doi.org/10.1039/b104827a]

[60] Kim, E.; Chufán, E.E.; Kamaraj, K.; Karlin, K.D. Synthetic models for heme-copper oxidases. *Chem. Rev.,* **2004**, *104*(2), 1077-1133.
[http://dx.doi.org/10.1021/cr0206162] [PMID: 14871150]

[61] Mahadevan, V.; Gebbink, R.K.; Stack, T.D.P. Biomimetic modeling of copper oxidase reactivity. *Curr. Opin. Chem. Biol.,* **2000**, *4*(2), 228-234.
[http://dx.doi.org/10.1016/S1367-5931(99)00080-0] [PMID: 10742191]

[62] Fontecave, M. Pierre, Oxidations by copper metalloenzymes and some biomimetic approachesJ. *Coord. Chem. Rev.,* **1998**, *170*, 125-140.
[http://dx.doi.org/10.1016/S0010-8545(98)00068-X]

[63] Zhang, C.X.; Liang, H.; Humphreys, K.J.; Karlin, K.D. *Advances in Catalytic Activation of Dioxygen*

by Metal Complexes; , **2003**.

[64] Baglia, R.A.; Zaragoza, J.P.T.; Goldberg, D.P. Biomimetic Reactivity of Oxygen-Derived Manganese and Iron Porphyrinoid Complexes. *Chem. Rev.,* **2017**, *117*(21), 13320-13352.
[http://dx.doi.org/10.1021/acs.chemrev.7b00180] [PMID: 28991451]

[65] Que, L., Jr; Tolman, W.B. Biologically inspired oxidation catalysis. *Nature,* **2008**, *455*(7211), 333-340.
[http://dx.doi.org/10.1038/nature07371] [PMID: 18800132]

[66] Otsuka, S.; Nakamura, A.; Tatsuno, Y. Oxygen complexes of nickel and palladium. Formation, structure, and reactivities. *J. Am. Chem. Soc.,* **1969**, *91*, 6994-6999.
[http://dx.doi.org/10.1021/ja01053a017]

[67] Fujita, K.; Schenker, R.; Gu, W.; Brunold, T.C.; Cramer, S.P.; Riordan, C.G. A monomeric nickel-dioxygen adduct derived from a nickel(I) complex and O2. *Inorg. Chem.,* **2004**, *43*(11), 3324-3326.
[http://dx.doi.org/10.1021/ic049876n] [PMID: 15154789]

[68] Yao, S.; Bill, E.; Milsmann, C.; Wieghardt, K.; Driess, M. A "side-on" superoxonickel complex [LNi(O2)] with a square-planar tetracoordinate nickel(II) center and its conversion into [LNi(mu-OH)2NiL]. *Angew. Chem. Int. Ed. Engl.,* **2008**, *47*(37), 7110-7113. [LNi(mu-OH)2NiL].
[http://dx.doi.org/10.1002/anie.200802234] [PMID: 18671225]

[69] Company, A.; Yao, S.; Ray, K.; Driess, M. Dioxygenase-like reactivity of an isolable superoxo-nickel(II) complex. *Chemistry,* **2010**, *16*(31), 9669-9675.
[http://dx.doi.org/10.1002/chem.201001138] [PMID: 20645352]

[70] Corona, T.; Pfaff, F.F.; Acuña-Parés, F.; Draksharapu, A.; Whiteoak, C.J.; Martin-Diaconescu, V.; Lloret-Fillol, J.; Browne, W.R.; Ray, K.; Company, A. Reactivity of a Nickel(II) Bis(amidate) Complex with meta-Chloroperbenzoic Acid: Formation of a Potent Oxidizing Species. *Chemistry,* **2015**, *21*(42), 15029-15038.
[http://dx.doi.org/10.1002/chem.201501841] [PMID: 26311073]

[71] Schröder, D.; Schwarz, H. C-H and C-C Bond Activation by Bare Transition-Metal Oxide Cations in the Gas Phase. *Angew. Chem. Int. Ed. Engl.,* **1995**, *34*, 1973-1995.
[http://dx.doi.org/10.1002/anie.199519731]

[72] Ryan, M.F.; Fiedler, A.; Schröder, D.; Schwarz, H. adical-like behavior of manganese oxide cation in its gas-phase reactions with dihydrogen and alkanes. *J. Am. Chem. Soc.,* **1995**, *117*, 2033-2040.
[http://dx.doi.org/10.1021/ja00112a017]

[73] Harvey, J.N.; Diefenbach, M.; Schröder, D.; Schwarz, H. Oxidation properties of the early transition-metal dioxide cations MO_2^+ (M = Ti, V, Zr, Nb) in the gas-phase. *Int. J. Mass Spectrom.,* **1999**, *182/183*, 85-97.
[http://dx.doi.org/10.1016/S1387-3806(98)14241-0]

[74] Schröder, D.; Schwarz, H. *Organometallic Oxidation Catalysis*; Springer: Heidelberg, **2007**, 22, pp. 1-15.

[75] Shiota, Y.; Yoshizawa, K. Methane−Methanol Conversion by MnO^+, FeO^+, and CoO^+: A Theoretical Study of Catalytic Selectivity. *J. Am. Chem. Soc.,* **2000**, *122*, 12317-12326.
[http://dx.doi.org/10.1021/ja0017965]

[76] Itoh, S.; Bandoh, H.; Nakagawa, M.; Nagatomo, S.; Kitagawa, T.; Karlin, K.D.; Fukuzumi, S. Formation, characterization, and reactivity of bis(μ-oxo)dinickel(III) complexes supported by a series of bis[2-(2-pyridyl)ethyl]amine ligands. *J. Am. Chem. Soc.,* **2001**, *123*(45), 11168-11178.
[http://dx.doi.org/10.1021/ja0104094] [PMID: 11697960]

[77] Shiren, K.; Ogo, S.; Fujinami, S.; Hayashi, H.; Suzuki, M.; Uehara, A.; Watanabe, Y.; Moro-oka, Y. Efficient and recyclable monomeric and dendritic Ru-based metathesis catalysts. *J. Am. Chem. Soc.,* **2000**, *122*, 254-262.
[http://dx.doi.org/10.1021/ja990311d]

[78]   Cho, J.; Furutachi, H.; Fujinami, S.; Suzuki, M. A bis(μ-alkylperoxo)dinickel(II) complex as a reaction intermediate for the oxidation of the methyl groups of the Me($_2$)-tpa ligand to carboxylate and alkoxide ligands. *Angew. Chem. Int. Ed.,* **2004**, *43*(25), 3300-3303.
       [http://dx.doi.org/10.1002/anie.200353637] [PMID: 15213958]

[79]   Cho, J.; Sarangi, R.; Annaraj, J.; Kim, S.Y.; Kubo, M.; Ogura, T.; Solomon, E.I.; Nam, W. Geometric and electronic structure and reactivity of a mononuclear "side-on" nickel(III)-peroxo complex. *Nat. Chem.,* **2009**, *1*(7), 568-572.
       [http://dx.doi.org/10.1038/nchem.366] [PMID: 20711413]

[80]   Cho, J.; Kang, H.Y.; Liu, L.V.; Sarangi, R.; Solomon, E.I.; Nam, W. Mononuclear nickel(II)-superoxo and nickel(III)-peroxo complexes bearing a common macrocyclic TMC ligand. *Chem. Sci. (Camb.),* **2013**, *4*(4), 1502-1508.
       [http://dx.doi.org/10.1039/c3sc22173c] [PMID: 23662168]

[81]   Nagataki, T.; Tachi, Y.; Itoh, S. Ni^{II}(TPA) as an efficient catalyst for alkane hydroxylation with *m*-CPBA. *Chem. Commun. (Camb.),* **2006**, (38), 4016-4018.
       [http://dx.doi.org/10.1039/b608311k] [PMID: 17003884]

[82]   Nagataki, T.; Ishii, K.; Tachi, Y.; Itoh, S. Ligand effects on Ni(^{II})-catalysed alkane-hydroxylation with *m*-CPBA. *Dalton Trans.,* **2007**, (11), 1120-1128.
       [http://dx.doi.org/10.1039/b615503k] [PMID: 17339995]

[83]   Balamurugan, M.; Mayilmurugan, R.; Suresh, E.; Palaniandavar, M. Nickel(II) complexes of tripodal 4N ligands as catalysts for alkane oxidation using *m*-CPBA as oxidant: ligand stereoelectronic effects on catalysis. *Dalton Trans.,* **2011**, *40*(37), 9413-9424.
       [http://dx.doi.org/10.1039/c1dt10902b] [PMID: 21850329]

[84]   Honda, K.; Cho, J.; Matsumoto, T.; Roh, J.; Furutachi, H.; Tosha, T.; Kubo, M.; Fujinami, S.; Ogura, T.; Kitagawa, T.; Suzuki, M. Oxidation reactivity of bis(μ-oxo) dinickel(III) complexes: arene hydroxylation of the supporting ligand. *Angew. Chem. Int. Ed. Engl.,* **2009**, *48*(18), 3304-3307.
       [http://dx.doi.org/10.1002/anie.200900222] [PMID: 19347913]

[85]   Pirovano, P.; Farquhar, E.R.; Swart, M.; Fitzpatrick, A.J.; Morgan, G.G.; McDonald, A.R. Characterization and reactivity of a terminal nickel(III)-oxygen adduct. *Chemistry,* **2015**, *21*(9), 3785-3790.
       [http://dx.doi.org/10.1002/chem.201406485] [PMID: 25612563]

[86]   Pirovano, P.; Farquhar, E.R.; Swart, M.; McDonald, A.R. Tuning the reactivity of terminal Nickel(III)–oxygen adducts for C–H bond activation. *J. Am. Chem. Soc.,* **2016**, *138*(43), 14362-14370.
       [http://dx.doi.org/10.1021/jacs.6b08406] [PMID: 27739688]

[87]   Pirovano, P.; Berry, A.R.; Swart, M.; McDonald, A.R. Indirect evidence for a Ni^{III}-oxyl oxidant in the reaction of a Ni^{II} complex with peracid. *Dalton Trans.,* **2017**, *47*(1), 246-250.
       [http://dx.doi.org/10.1039/C7DT03316H] [PMID: 29211082]

[88]   Pirovano, P.; Twamley, B.; McDonald, A.R. Modulation of nickel pyridinedicarboxamidate complexes to explore the properties of high☐valent oxidants. *Chemistry,* **2018**, *24*(20), 5238-5245.
       [http://dx.doi.org/10.1002/chem.201704618] [PMID: 29206304]

[89]   McManus, C.; Mondal, P.; Lovisari, M.; Twamley, B.; McDonald, A.R. Carboxamide ligand noninnocence in proton coupled electron transfer. *Inorg. Chem.,* **2019**, *58*(7), 4515-4523.
       [http://dx.doi.org/10.1021/acs.inorgchem.9b00055] [PMID: 30864788]

[90]   Bunescu, A.; Lee, S.; Li, Q.; Hartwig, J.F. Catalytic hydroxylation of polyethylenes. *ACS Cent. Sci.,* **2017**, *3*(8), 895-903.
       [http://dx.doi.org/10.1021/acscentsci.7b00255] [PMID: 28852704]

[91]   Qiu, Y.; Hartwig, J.F. Mechanism of Ni-catalyzed oxidations of unactivated C(sp^3)–H bonds. *J. Am. Chem. Soc.,* **2020**, *142*(45), 19239-19248.
       [http://dx.doi.org/10.1021/jacs.0c09157] [PMID: 33111517]

[92]   Rajpurohit, J.; Shukla, P.; Kumar, P.; Das, C.; Vaidya, S.; Sundararajan, M.; Shanmugam, M.; Shanmugam, M. Stabilizing Terminal Ni(III)–hydroxide complex using NNN-Pincer ligands: Synthesis and characterization. *Inorg. Chem.,* **2019**, *58*(9), 6257-6267.
[http://dx.doi.org/10.1021/acs.inorgchem.9b00466] [PMID: 31009214]

[93]   Pfaff, F.F.; Heims, F.; Kundu, S.; Mebs, S.; Ray, K. Spectroscopic capture and reactivity of $S = 1/2$ nickel(III)-oxygen intermediates in the reaction of a Ni(II)-salt with *m*CPBA. *Chem. Commun. (Camb.),* **2012**, *48*(31), 3730-3732.
[http://dx.doi.org/10.1039/c2cc30716b] [PMID: 22398975]

[94]   Pandey, B.; Ray, K.; Rajaraman, G.Z. Structure, bonding, reactivity and spectral features of putative Ni^{III}=O species: A theoretical perspective. *Anorg. Allg. Chem.,* **2018**, *644*, 790-800.
[http://dx.doi.org/10.1002/zaac.201800122]

[95]   Hikichi, S.; Hanaue, K.; Fujimura, T.; Okuda, H.; Nakazawa, J.; Ohzu, Y.; Kobayashi, C.; Akita, M. Characterization of nickel(II)-acylperoxo species relevant to catalytic alkane hydroxylation by nickel complex with *m*CPBA. *Dalton Trans.,* **2013**, *42*(10), 3346-3356.
[http://dx.doi.org/10.1039/C2DT32419A] [PMID: 23223606]

[96]   Hikichi, S.; Okuda, H.; Ohzu, Y.; Akita, M. Structural characterization and oxidation activity of a nickel(II) alkylperoxo complex. *Angew. Chem. Int. Ed. Engl.,* **2009**, *48*(1), 188-191.
[http://dx.doi.org/10.1002/anie.200804402] [PMID: 19040236]

[97]   Nakazawa, J.; Terada, S.; Yamada, M.; Hikichi, S. Structural characterization and oxidation reactivity of a nickel(II) acylperoxo complex. *J. Am. Chem. Soc.,* **2013**, *135*(16), 6010-6013.
[http://dx.doi.org/10.1021/ja4020277] [PMID: 23582051]

CHAPTER 4

Mechanisms of Some Heterogeneous Photocatalytic Reactions of Oxidation Occurring *via* Oxygen Atom Transfer

Robert Bakhtchadjian[1,*]

[1] Institute of Chemical Physics of the National Academy of Sciences of the Republic of Armenia, Yerevan, Republic of Armenia

Abstract: Insights into the mechanisms of oxygen atom transfer in the photooxidation of organic compounds for heterogeneous photocatalytic systems have been presented. These reactions have a wide variety of practical applications in chemistry, biology, and applied sciences. The role of the oxygen atom transfer mechanism in homogeneous photocatalysis has been investigated for nearly a century. Relatively little attention has been paid to the disclosure of oxygen atom transfer reactions in heterogeneous photocatalytic systems. This chapter discusses some problems related to the catalytic oxygen atom transfer in the oxidation of organic compounds, mainly with dioxygen, under UV irradiation or visible light, in heterogeneous reaction systems. Various active oxygen species, including oxygen atom transfer agents, as reaction intermediates can be generated in these systems. Depending on the nature of the active oxygen species, including metal-oxo compounds, the photoassisted catalytic oxygen atom transfer can occur mainly by the primary photoexcitation of either the catalyst or its photosensitive solid support, in rare cases, also the catalyst/support complex. The peculiarities of the mechanism of photo-driven oxygen atom transfer were mainly exemplified by the reactions occurring in heterogeneous catalytic systems containing transition metal oxides, their metalorganic complexes, and other photosensitive solid materials, including heterogenized homogeneous photocatalysts on the different supports, such as the transition metal-oxo complexes on the semiconductor materials. Special attention has been paid to the chemistry of TiO_2 and TiO_2-based semiconductor photocatalysis from the point of view of the reaction mechanisms, including oxygen atom transfer reactions.

Keywords: Oxygen atom transfer, Heterogeneous photocatalytic reactions, Reaction mechanism.

---
* **Corresponding author Robert Bakhtchadjian**: Institute of Chemical Physics of the National Academy of Sciences of the Republic of Armenia, Yerevan, Republic of Armenia ; E-mail address: robakh@hotmail.com

INTRODUCTION

Photo-driven catalytic reactions are divided into two main branches: homogeneous and heterogeneous photocatalysis [1, 2], although each of them can often be expressed as a bimodal reaction sequence when the key reaction intermediates are transferred from one phase to another [3]. Both heterogeneous and homogeneous photochemical oxidation of organic compounds with dioxygen or other oxidants, more often called photoredox reactions, as a rule, are complex, multistage, and sometimes even multiphase processes occurring *via* the formation and further reactions of different intermediates [4].

The mechanism of photoredox reactions is principally different from the mechanism of thermal reactions, although sometimes the reaction intermediates can be the same in both cases. In a photochemical system that absorbs light energy, one or more components are transferred into excited states, and their further interactions lead to the formation of reaction intermediates and/or products. In this regard, photoredox systems can produce such intermediates (by relaxation of the energetically excited intermediates from a higher energy state to a lower energy state), which may not be generated during thermal oxidation due to energy inaccessibility [5].

Homogeneous photochemical reactions of the oxidation of organic substances, particularly, with dioxygen, in the absence of metal ions or organometallic complexes, are usually radical and chain, often autocatalytic processes. In the presence of metal ions in homogeneous reaction media, the predominant mechanisms of photochemical oxidation are similar to those of reactions with the Fenton reagents, known as photo-Fenton reactions [6]. In the case of the participation of transition metal ions or their organometallic complexes catalyzing oxidation reactions under UV irradiation or visible light, the transfer of electrons or atoms depends mainly on the nature of the central metal ions and their coordination environment [7]. Oxygen atom transfer occurs either by direct insertion of the oxygen atom into the substrate or by the so-called oxygen rebound (radical) mechanisms, *via* the primary abstraction of hydrogen atom from the substrate. The transfer of oxygen atoms can occur either as a stepwise or as a concerted reaction [7].

In heterogeneous photocatalysis, the light energy is absorbed by the solid phase photosensitive component(s) of the chemical system that transforms it into the chemical bond energy. If this transformation takes place on the surface of the catalyst, the photocatalytic process is named "direct", and, if it occurs through the absorption of light energy by the substrate or substrate-catalyst complex, as well as by another reaction component capable of transferring it to the catalyst or substrate, it is named "indirect" [8]. The majority of the heterogeneous photocatalysts are semiconductors (TiO_2, CuO, ZnO, NiO), dyes, organic or inorganic polymeric materials [2, 4, 9]. Among them, the exceptional usefulness of the photocatalytic application of "bare" and surface modified TiO_2 has been revealed in a large number of investigations in recent decades. Taking into consideration the wide application and practical importance of TiO_2 and TiO_2-based photocatalysts in wastewater treatment (AOP technologies), elimination of certain atmospheric pollutants, disinfection of surfaces, *etc* [6, 9, 10], this chapter is mainly devoted to the problems of the mechanism of oxidation in these and analogous systems. Here, we will discuss mainly the problems of "direct" photocatalytic reactions [8].

Investigations of the oxygen atom transfer in heterogeneous photocatalytic reactions are related to the determination or estimation of certain kinetic parameters (rate constant, quantum yield, turnover number, photonic efficiency), the accurate measurement of which is often complicated caused by the complexity of the multiphase and multicomponent photochemical systems. In this regard, it will be shown that the existing data often provide limited possibilities to make summarizing conclusions about the prevalence of one or other mechanisms in oxidation *via* oxygen atom transfer reactions.

Special attention will be paid to heterogeneous photocatalytic systems, where the transfer of oxygen atoms to substrates occurs from transition metal-oxo complexes anchored on the surfaces of semiconductors or supported by non-photosensitive materials. The revealing of the mechanisms of oxo-atom transfer from transition metalorganic complexes may serve as a functional model for enzymatic oxidation, as well as a key to the creation of new photocatalytic redox systems, corresponding to the requirements of "green" and sustainable chemistry.

Photocatalytic Oxidation Reactions Occurring *via* Oxygen Atom Transfer

Heterogeneous photocatalytic oxidation reactions of organic compounds on the surfaces of metals, metal oxides, organometallic complexes, and other photosensitive solid substances occur through the formation of intermediates of different nature (radicals, ions, excited species, metastable compounds, *etc.*) and, therefore, in general, they have different reaction mechanisms. For example, according to the more or less generally accepted mechanisms for the oxidation of alcohols to aldehydes with dioxygen on the noble metals, usually, the main intermediates are alkoxy species adsorbed on the surface of the catalyst, the dehydrogenation of which leads to the formation of aldehydes (Scheme **1**) [11].

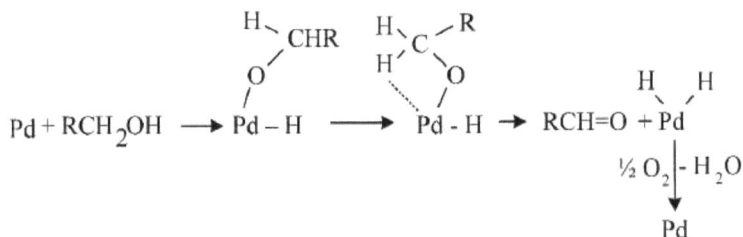

Scheme (1). Pd-catalyzed oxidation of alcohols with dioxygen [11].

Here, the essential role of the oxidizing agent, dioxygen, is the regeneration of the catalyst by the dehydrogenation reaction that produces water.

Other analogous examples of the mechanisms involving the metal-alcoholate species, and β-hydride elimination from the alcoholate, are Ru/Al$_2$O$_3$ catalyzed oxidation of alcohols, diols, and amines [12].

Completely different mechanisms are predominant in the case of the transition metal oxides in photochemical oxidation with dioxygen. Particularly, in the photocatalytic oxidation of organic compounds on TiO$_2$ with dioxygen, the most accepted mechanism is related to the formation of the active (or reactive) oxygen species as the main intermediates. They consist of a great number of chemical entities of different classes: oxygen radicals (O, O$^-$, O$_2^-$), singlet oxygen (^{1}O$_2$), and radicals or molecules containing not only oxygen but also a number of other elements (OH, HO$_2$, NO, CO$_2^-$, RO$_2$, H$_2$O$_2$, organic peroxide compounds), including also metal(M)-oxygen moieties, as M=O^{2-} or M-O^{2-}-M [11, 13]. Active oxygen species, such as O, O$^-$, O$_2^-$, O$_2^{2-}$, usually exhibit electrophilic properties, while the terminal and bridging metal-oxygen species, in general, are nucleophilic or amphoteric [13]. All of these and a great number of other species formed as a result of photocatalytic reactions on the surfaces of solid substances are candidates for key intermediates in the oxidation of organic compounds, depending on the reaction conditions in the chemical system. As oxidant agents, they exhibit different reactivity in the fluid phases and in the adsorbed state on the surfaces of solid substances [3]. The adsorbed species on different surfaces, which are free radicals in the fluid phases, are often classified as radical-like species [3]. Among the above-mentioned oxygen active species, the chemical entities that include oxygen atom(s) combined with a metal element(s) and compose moieties of different inorganic or organometallic compounds (metal oxides; metal-oxo, oxyl, or peroxo complexes) are of particular interest in both chemistry and biology, in homogeneous and heterogeneous photocatalysis, as oxygen atom transfer agents [7, 11, 13].

The primary action of light on the surface layer of semiconductor photocatalyst ($\lambda > 360$ nm for TiO_2), called photoexcitation, leads to the separation of charge carrier pairs: h^+ (holes in valence band) and e^- (electrons in conductive band) [2, 4, 5, 7]. There are at least three possible pathways for further interactions of holes and electrons: recombination; trapping in defect sites in the bulk and on the surface of solid substances; and interactions with chemical compounds by charge transfer reactions. It is obvious that only the third pathway is significant in photochemistry, because two other pathways are chemically unproductive. For example, the generation of radicals $OH^{\bullet}$ and $O_2^{\bullet-}$ on TiO_2 in aqueous media and in the presence of dioxygen may occur *via* the following interactions:

$$h^+ + OH^- \rightarrow OH^{\bullet}$$

$$e^- + O_2 \rightarrow O_2^{\bullet-}$$

The formation of the active oxygen species and their role in oxidation on TiO_2, under UV irradiation are discussed in nearly every modern review and handbook devoted to the problems associated with the photocatalytic redox reactions. In most mechanisms of the oxidation of organic compounds on the surfaces of TiO_2, it was considered that the actual oxidants were mainly holes h^+, radicals $OH^{\bullet}$, $O_2^{\bullet-}$, $HO_2^{\bullet}$ or H_2O_2 and organic peroxides. In principle, they may continue the reaction through different pathways, including:

(i)-generation of other radicals, radical like or other surface-active species; decomposition and/or transfer of them from the surface to the fluid phases
(ii)-reaction with the substrate and the formation of final products *via* the transfer of electrons; transfer of oxygen atom; abstraction of hydrogen atom, *etc.*

Note that the photoexcitation or primary action of light at the surface level of semiconductors is essentially different from that in the case of the surface metal-oxo complexes supported by different materials. If the support is not a photosensitive material, under appropriate conditions, part of the UV irradiation or light energy may be absorbed by the metal-oxo complex and used to transfer electron(s) from the metal to oxygen with the formation of $M^+\text{-}O^-$ (metal oxyl) intermediates. The photoexcitation of semiconductor materials used as supports of metal-oxo complexes in photocatalytic reactions is discussed in the following section.

It is also important to note an essential peculiarity observed in many heterogeneous photocatalytic reactions that can often remarkably change the perceptions of the progression of photochemical processes after the formation of radicals or radical-like species on the surfaces of semiconductor catalysts, particularly on TiO_2. In the early stages of the development of heterogeneous photocatalysis, the mechanisms of oxidation reactions were usually related only to the primary chemical reactions on the surfaces of semiconductors under the influence of light or UV irradiation. For example, in the "classical" photocatalytic reaction, the splitting of water on a semiconductor electrode (TiO_2, in the presence of Pt-electrode), the following "purely" heterogeneous reactions on TiO_2 have been suggested [8].

TiO_2 electrode $\qquad (TiO_2) + h\nu \rightarrow (TiO_2)(e^- + h^+)$

$$2H_2O + 4h^+ \rightarrow O_2 + 4H^+$$

Pt electrode $\qquad 2H^+ + 2e^- \rightarrow H_2$

And the summary reaction is:

$$2H_2O + 4 h\nu \xrightarrow{TiO_2} 2H_2 + O_2$$

However, the accumulation of experimental data on the oxidation of organic compounds driven by UV irradiation on TiO_2 revealed the impossibility of accurately describing these reactions within the framework of models of "purely" heterogeneous processes. For example, it was clearly shown experimentally that the photogenerated OH radicals (reaction of holes with water or hydroxyl ions), which were considered key intermediaries in many photocatalytic oxidation or oxidative destruction reactions of organic compounds on the TiO_2 surface, can be of at least two types: surface-bonded hydroxyl OH_s and surface free hydroxyl OH_f [3 (chapter 4), 15, 16]. These two types of intermediates have very different reactivity on the surface. Moreover, a great number of investigations have confirmed that a part of OH_f radicals can be desorbed from the surface and diffused to the bulk of the fluid phases. It has been considered that OH_s is bonded with only one Ti-atom, and OH_f with two Ti-atoms [16 - 18]. The OH radicals appeared in the fluid phases are capable of continuing the oxidation by the radical pathway. This transfer of radicals from the solid surface to the gas or liquid phases was confirmed by data obtained using a number of modern experimental methods, such as laser induced fluorescence spectroscopy [17], EPR-method [19], ATR-IR (attenuated total reflection in infrared spectroscopy) [16], CRDS (cavity ring down spectroscopy) [20]. The brief review of the heterogeneous-homogeneous photocatalytic processes is presented in monograph [3]. All these experimental investigations clearly demonstrate that in a great number of heterogeneous photocatalytic oxidation reactions, the overall reaction can be described only by taking into consideration the bimodal reaction sequence [3, 21]. A nearly analogous conclusion was drawn by the authors of the work [8], stating that *"some photocatalytic reactions have been reported wherein the photocatalytically produced compounds are not the final products. In fact, they react in the solution bulk or catalytically on the surface of the semiconductor with other species, thus producing the target compounds"*. Therefore, *"investigating and proposing mechanistic insights, the concepts of photocatalysis and catalysis or reactions in bulk solution or adsorbed phase may often overlap and it is useful to analyze them according to an interdisciplinary over-view"*.

Returning to the oxygen atom transfer reactions in understanding the mechanism in photocatalytic oxidation of organic compounds on TiO_2 surfaces, note that they were revealed in a number of investigations [22 - 29]. Zhao and coworkers [23], applying the isotope labeled technique, clearly showed the transfer of oxygen atoms in photocatalytic oxidation of organic compounds on the TiO_2 surface, for example, *via* oxygen atom transfer from O_2 to alpha-carbon of alcohol forming aldehydes in benzotrifluoride (BTF) solution. The predominant overall reaction has been presented as follows:

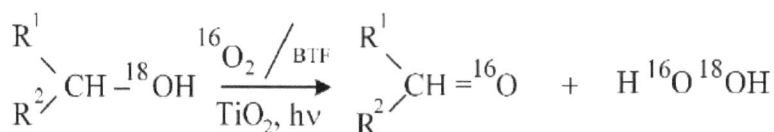

The following reaction scheme (Scheme **2**) demonstrates the suggested mechanism of oxygen atom transfer in photocatalytic oxidation of alcohols with dioxygen [23].

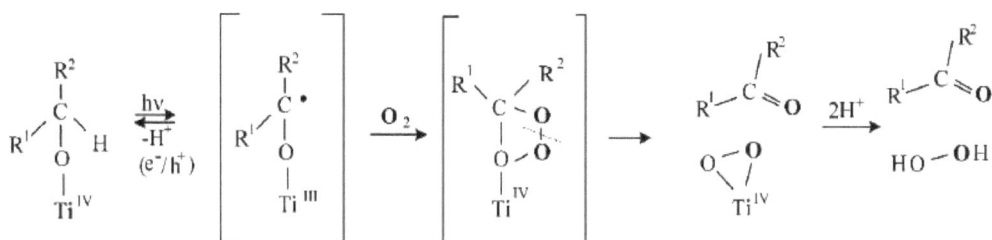

Scheme (2). Oxygen atom transfer in the photocatalytic oxidation of alcohols with dioxygen on a TiO_2 [23].

According to the perceptions developed by Zhao *et al.* [23 - 25], in the case of TiO_2, the electron transfer to the substrate is an interfacial process between the bulk of the liquid phase and the surface of the solid phase. Particularly, this concept has been experimentally evidenced by the application of the mentioned isotope labeled technique, when the solvent was water [24]. The following example demonstrates a peculiarity of the hydroxylation of the aromatic ring with dioxygen on the surface of TiO_2 in a water medium (Fig.1):

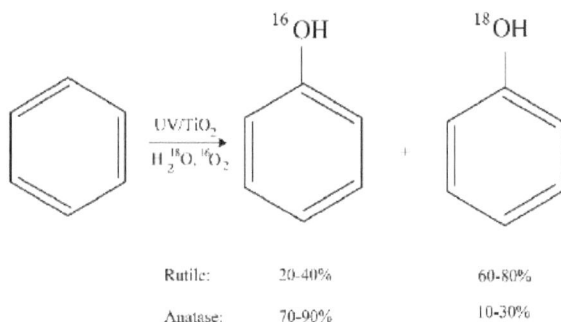

	^{16}OH	^{18}OH
Rutile:	20-40%	60-80%
Anatase:	70-90%	10-30%

Fig. (1). Photocatalytic oxidation on TiO_2 of benzene to phenol in aerated aqueous solution (Adapted from Ref [24].).

Zhao et al. [24] also investigated the photocatalytic reactions of hydroxylation of other aromatics, oxidative cleavage of aryl rings, and photocatalytic decarboxylation of saturated carboxylic acids on the TiO_2 surface in the presence of dioxygen. According to the authors [24], the activation of oxygen in hydroxylation occurs by the interaction of dioxygen with conduction band electrons. In the oxidative cleavage reaction, the insertion of O-atoms takes place through the so-called "site-dependent coordination of reactants". In the decarboxylation of acids, O_2 is activated by conduction band electrons and incorporated into the pyruvic acid intermediate.

All of the above results confirm the general conclusion that oxygen atom transfer in heterogeneous photochemical oxidation or oxidative decomposition (destruction) reactions with dioxygen on semiconductors is one of the main pathways in the mechanism [28, 29].

About Two Classes of Heterogeneous Photocatalytic Reactions: Photogenerated Catalysis and Catalyzed Photolysis

In classifying the photocatalytic reactions, Salomon differentiated two main classes of these processes [30, 31]:

1. Photogenerated (photoinduced, photoinitiated), when the light-generated catalyst is in a ground state electron configuration and it "interacts with the substrate to carry out the thermodynamically spontaneous catalytic step" [30], (Fig. (**2a**). It is also known as true catalysis.

2. Catalyzed photolysis (photosensitized), ("Catalyzed photolysis" is a term that is currently not widely used), when the molecular entity (substrate or reactant) absorbs light and "induces a chemical and physical alteration of another chemical entity (photosentizer, catalyst)." During the catalyzed photolysis, the catalyst or substrate, or the catalyst and substrate together, are in an electronically excited state (Fig. (**2b**) [31].

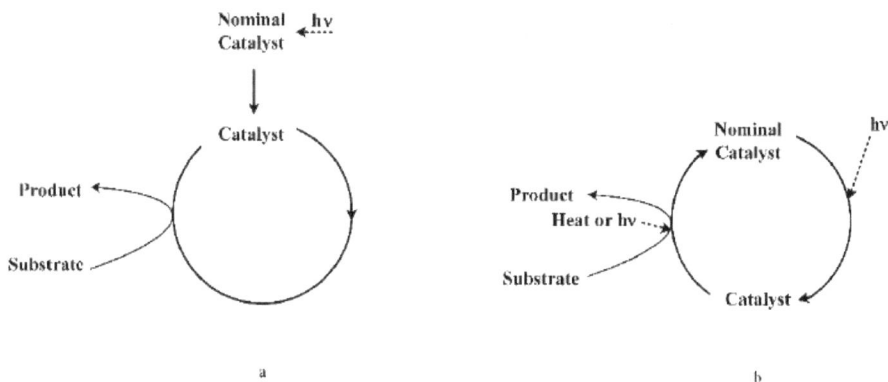

Fig. (2). Schemes of the catalytic cycles in (**a**) photogenerated catalysis; (**b**) catalyzed photolysis; according to Serpone and Emeline (Adapted from the work [31]).

In a homogeneous photogenerated (photoinduced) catalytic reaction, the precursor of a catalyst or a nominal catalyst that absorbs and transforms light energy, becomes an electronically (chemically) modified real-catalyst that carries out chemical transformations. The photoinduced formation of free radicals or ions capable of initiating a chain reaction can be considered a particular case of the photogenerated reaction [32]. Theoretically, one or two photons may generate either radicals, which, in turn, can develop chain reactions, or other reactive species performing catalytic cycles with a high turnover number. In the case of the catalyzed photolysis (photosensitized), as well as photoassisted reactions in homogeneous media, photons are the initial quasi-reactants participating in every catalytic cycle of the chemical transformation. The two main categories of photocatalytic homogeneous processes have also been attributed to the heterogeneous photoredox reactions, and have been named: heterogeneous photogenerated (photoinitiated, photoinduced) and heterogeneous catalytic photolysis (photosensitized, photoassisted) processes [31, 33]. It should be noted the principal differences between the two mentioned classes of heterogeneous photocatalytic reactions, as the activation mechanism of the catalysts and kinetic parameters of the systems are different. As mentioned in [32], in this case, "we deal with different catalysts". In the case of the catalyzed photolysis, under irradiation, the Fermi level of the semiconductor (catalyst) is split into two quasi-Fermi levels. Contrary, in the case of the photogenerated catalysis, the Fermi level is unique, although it is shifted in comparison with the initial state, and the reaction may occur in the dark after the preexcitation of the nominal catalyst [32].

The mentioned classes of photochemical reactions, in general, can be differentiated by measuring or estimating values of the *quantum yield of reactions* (Φ). In homogeneous or heterogeneous reactions, the quantum yield is the following ratio [31]:

$$\Phi = \Delta(substrate)/\ number\ of\ photons\ absorbed\ by\ system$$

Δ(substrate) is the number of consumed molecules of the substrate. Obviously, Φ can be compared per photon and per unit time, and in a heterogeneous system, as well per unit surface. Other expressions of Φ are given in [31 - 33]. Certain considerations related to the definition of the quantum yield in multiphoton elementary processes were discussed in [34]. In a homogeneous photoinduced chain-radical reaction $\Phi \gg 1$, as in the case of the reaction of H_2 with Cl_2, $\Phi \gg 1$ (800-3000) in the gas phase. In a homogeneous photoassisted (photosensitized) reaction, the values of the quantum yield or quantum efficiency of reaction may be $\Phi \leq 1$ or $\leq 100\%$ [35]. Unlike homogeneous systems, the experimental determination of quantum yield in heterogeneous photocatalytic reactions is more complicated. Since the 1990s, it has remained a subject of much debate in the scientific literature [36 - 43]. Unfortunately, the values of the quantum yield and quantum efficiency in heterogeneous photocatalytic systems, according to Serpone [37], usually, "are ill-defined" and any reference to quantum yields is "ill-advised unless the actual number of photons absorbed by the light harvester (the photocatalyst) has been determined." In general, light scattering in heterogeneous photocatalytic systems is very significant and depends on different experimental conditions (light sources, reactor geometries, sizes of the suspended particles, *etc.*) [38]. The loss of photon flux as a result of the scattering and reflection of light by the suspended particles may be 13-76%, according to data [38]. In this regard, it has been proposed an alternative for the comparative determination of quantum yields by measuring the so-called apparent quantum yield or quantum efficiency, which is defined as a relative photonic efficiency ζ_r [38]. For instance, the apparent quantum yield can be determined from $\varphi = \zeta\varphi_{phenol}$, where φ_{phenol} is the quantum yield of the reaction of oxidative disappearance of phenol (as a standard secondary actinometer), in photocatalysis on TiO_2 (Degussa P-25), or using the so-called ISO-recommended oxidation of NO_x or acetaldehyde on TiO_2 [38]. IUPAC also recommends to use the apparent quantum yield to describe heterogeneous photocatalytic systems [39].

A method for the experimental determination of the quantum yield in a heterogeneous photocatalytic system was described by Machuca *et al.* in [40]. They estimated the quantum yield of the TiO_2-based photocatalytic oxidative degradation of dichloroacetic acid (DCA) in the so-called differential perfect-mixture reactor in the range of 275-580 nm.

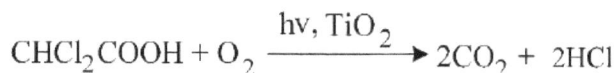

$$CHCl_2COOH + O_2 \xrightarrow{\text{hv, } TiO_2} 2CO_2 + 2HCl$$

The average value of the so-called "global" quantum yield for this photocatalytic system was 0.48 ± 0.20 [mol $\cdot$ einstein^{-1}]. Experimental estimation of the photonic efficiency and quantum yield of the formation of formaldehyde from methanol in the presence of various TiO_2 photocatalysts showed that they depend on the p^H of the suspension [41]. The maximum photonic efficiency was 10.4% and the quantum yield 0.08 when the absorbed photon flux was 4.9×10^{-8} Ein/L s. A comparative study of the kinetics of triphenylphosphine oxidation reactions with three compounds: $[Mo(VI)O_2X_2]L$ (complex compound, where X and L are ligands); H_2MoO_4; and MoO_3, anchored on the surfaces of TiO_2 [43], under UV irradiation (λ=360 nm) showed that only the bipyridine dioxo-Mo anchored complex provides the conversion of the initial reactant to triphenylphosphineoxide in stoichiometric quantities, with nearly 100% selectivity of the product. It was found that MoO_3 anchored on the surfaces of TiO_2 was inactive, and H_2MoO_4 was less active than the dioxo-Mo(VI) complex, in more or less comparable conditions of these reactions. An estimation of the quantum yield (Φ) for the reaction of the anchored complex showed that $\Phi<1$ (0.67 Ein s^{-1}), and the increase in the reaction rate correlated with the photonic flux to the surface. These data indicate that the overall photochemical reaction can be characterized as a photoassisted process occurring *via* oxo-atom transfer from the Mo-center, which is facilitated by the electronic flux from photosensitive TiO_2 to the metallic center of the anchored complex through the covalently bonded ligand with the surface.

In heterogeneous photocatalytic redox reactions on TiO_2, the estimation of the quantum yields (or analogues parameters) shows that usually $\Phi<1$. This indicates a photoassisted heterogeneous catalytic reaction. Regarding the photoinitiated (photogenerated) reactions with a possible chain-radical mechanism, it can be noted that theoretically, the expected values of the quantum yields, in this case, must be greater than unity. To the best of our knowledge, there are no data in the scientific literature concerning heterogeneous photocatalytic systems with the quantum yields (or apparent quantum yields) greater than unity or quantum efficiency greater than 100% [44]. However, there are experimentally well-proven examples of the photoinduced (photoinitiated) heterogeneous chain-radical reactions, such as hydrosilylation of hexene, benzaldehyde and allylamine on silicon surface, photoinitiated by UV irradiation [45], oxidation of H-terminated silicon in the presence of O_2 and H_2O initiated by UV-irradiation [46], in atmospheric chemistry, the so-called particle-to-phase reactions for example, OH-initiated reactions of oxidation of NO to NO_2 on the aerosol surface [47]; and many others under sunlight or UV irradiation [3]. Apparently, the lack of quantum yield values for these and analogous photocatalytic processes is related to the experimental difficulties in determining them, as already mentioned above.

Thus, it seems that most of the known reactions of oxygen atom transfer are photoassisted or photosensitized processes, although today, the problem does not have an exhaustive answer. Finally, it should be noted that in the scientific literature, terms related to photochemistry are not always used precisely, in particular, they do not always correspond to the peculiarities of these two classes of photocatalytic processes and their reaction mechanisms [3 p.197, 8].

Oxygen Atom Transfer from Transition Metal Oxo-complexes in Heterogeneous Photocatalytic Redox Systems

Organometallic complexes containing transition metal-oxo moieties constitute an important part of compounds capable of transferring an oxygen atom in photoredox systems. They are becoming increasingly important in the chemical, biochemical, pharmaceutical and electronic industries [48]. A number of redox processes catalyzed by transition metal-oxo complexes, including photocatalytic reactions, in homogeneous conditions, often provide very high selectivity of the formation of target products, for instance, some reactions of the photooxidation of organic compounds catalyzed by porphyrins, reviewed in [49 a, 49 b]. However, the homogeneous catalytic reactions have a number of limitations on an industrial scale [3, 49b]. Some of these are related to the problems of the separation of catalysts from the reaction mixture which is economically disadvantageous. Another important limitation is related to the thermal regimes of reactions in solvents. In general, the majority of the solvents used on an industrial scale are stable in heating until 100-150°, in rare cases, even at high pressures, until about

200°C. Moreover, in a number of reactions, there are essential differences related to the lifetimes of certain key intermediates affecting the kinetics and yields of products. For instance, some metal-oxo complexes tend to oligomerization (or polymerization) under certain conditions. This will be shown below on the example of an oxo-Mo complex.

The limitations arising in homogeneous organometallic catalysis owing to the mechanism of oxygen atom transfer can be overcome by applying an appropriate heterogeneous reaction system [50]. Often, the use of the heterogeneous support adsorbing catalyst can remarkably improve the photocatalytic activity [50]. For example, an improved heterogeneous photocatalyst (FePc/SiO$_2$) in the degradation of phenol, in an aqueous medium can be obtained by the adsorption of iron(II) phthalocyanine (FePc) onto silica [51]. Heterogenization of homogeneous catalysts by immobilizing transition metal-oxo complexes on the surfaces of different synthetic and natural materials in the solid state *via* chemical grafting with covalent bonds is a very efficient approach for certain reactions [50]. One of the best examples among a great number of others, is the immobilization of Schiff base complexes of transition metal ions (Fe^{2+}, Co^{2+}, V^{2+}, Cu^{2+}) onto graphene oxide sheets by covalent bonds [52]. The anchored copper (II) Schiff base complex exhibited improved catalytic properties when styrene was oxidized to epoxide with BuOOH (94% conversion and 99% selectivity).

In [53], it was shown that the catalytic activity of transition metalorganic complexes by oxo-atom transfer mechanisms could be significantly increased due to the heterogenization of the reaction system and the subsequent change of the photoexcitation mechanisms. In the asymmetric epoxidation of *trans*-stilbene to epoxide with dioxygen in the presence of immobilized manganese porphyrin on graphene-oxide (GO-[Mn(T2PyP)(tart) (tart)], a nanocomposite involving a covalently bonded catalyst, also coordinating of the imidazole molecule with the metal center [53], in a reaction driven by visible light (LED light source) provides a high turnover number (approximately 3000) and selectivity (approximately 100%) [53]. It is interesting that the photochemical oxidation and isomerization reactions of stilbene in the presence of dioxygen were first discovered by Ciamician (one of the pioneers of modern organic photochemistry) and Silber, in the early years of the last century [54].

The oxygen atom transfer reaction of oxo-Mo complexes plays an essential role in understanding the biochemical mechanisms of many enzymatic processes [55]. Previously, the reactivity of terminal dioxo-molybdenum(VI) organometallic complexes, bearing different ligands, in photoassisted reactions towards arenes, aryl alkanes and alkenes, alcohols, phosphines and their derivatives, under UV irradiation, were shown in a great number of investigations, which were reviewed in [56, 57]. For example, the dioxo Mo-complexes [(O=Mo(VI)=O)X$_2$]L where L is an organic bipyridine ligand, and X = Cl, Br, SCN, exhibit catalytic activity through the transfer of an oxo-atom to the substrate in the presence of second oxidant agents, such as (CH$_3$)$_2$SO, H$_2$O$_2$, N$_2$O, O$_2$ (Fig. **3**) [56].

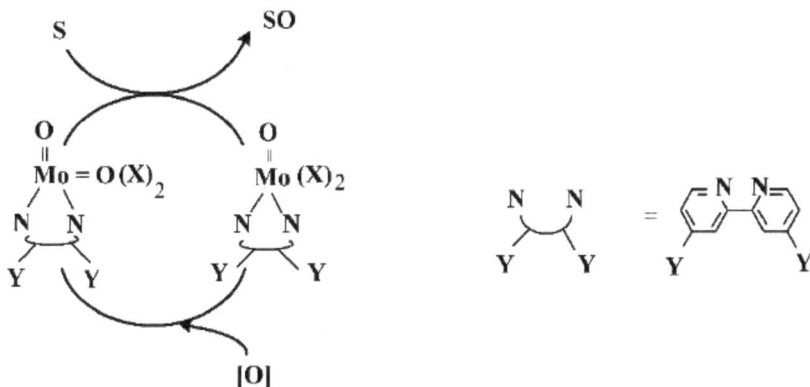

Fig. (3). Dioxo-Mo organometallic complexes in the catalytic cycle of the oxidation reaction, where X = Cl, Br, NCS; Y = CH$_3$, (CH$_3$)$_3$C, COOH, COOCH$_3$, N$_2$O, OCH$_3$, Cl; S = phosphine, alcohol, alkane, olefin; [O] = (CH$_3$)$_2$SO, H$_2$O$_2$, N$_2$O, O$_2$ S is a substrate [57].

The catalytic cycle in the presence of dioxygen consists of the following reactions *(I)-(III)*:

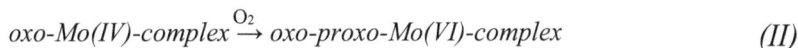

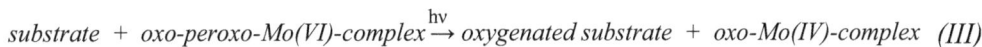

$$\text{substrate + dioxo-Mo(VI)-complex} \xrightarrow{h\nu} \text{oxygenated substrate + oxo-Mo(IV)-complex} \quad (I)$$

$$\text{oxo-Mo(IV)-complex} \xrightarrow{O_2} \text{oxo-proxo-Mo(VI)-complex} \quad (II)$$

$$\text{substrate + oxo-peroxo-Mo(VI)-complex} \xrightarrow{h\nu} \text{oxygenated substrate + oxo-Mo(IV)-complex} \quad (III)$$

Crystallographic analysis revealed that each of the mentioned dioxo-Mo(VI)-complexes also contained trace amounts of the μ-oxo (M-O-M) structures [58]. It appeared that μ-oxo (M-O-M) existed in the form of two conformers in the solution. Moreover, H^1 NMR-spectroscopic data clearly showed that in the solution, their quantities were in chemical equilibrium [58].

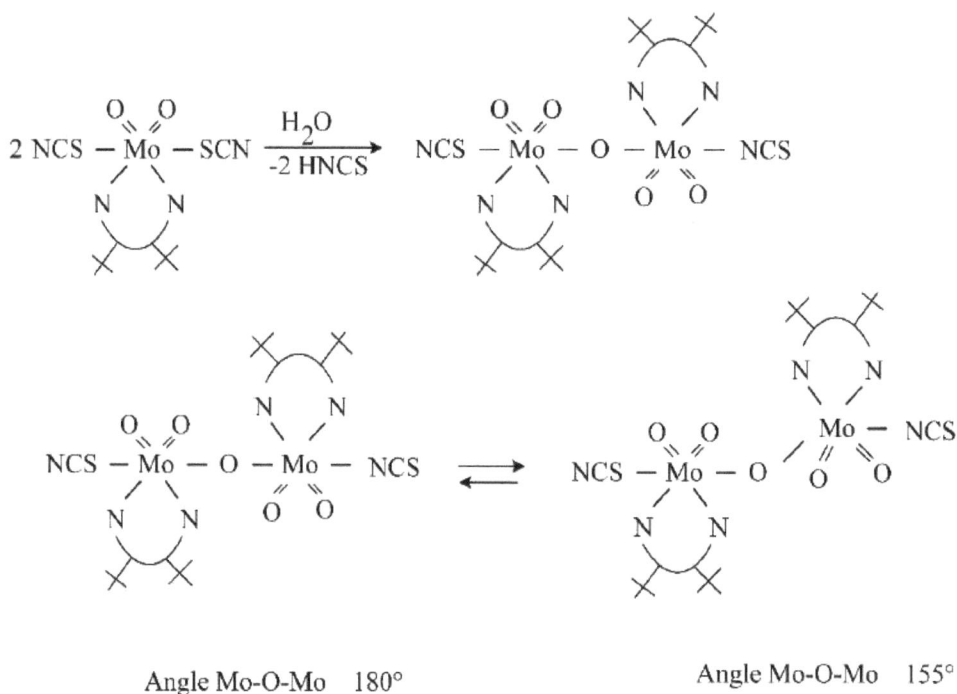

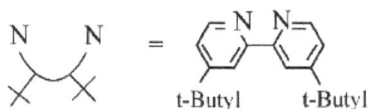

Scheme (3). Formation of dimer species from dioxo-Mo(VI) organometallic complex compound [58].

Both conformers exhibit oxidizing capacity towards many organic compounds. Further detailed investigation of the transfer of oxygen atom from these complexes, mainly by the method of UV spectroscopy accompanied by NMR- and X-ray-crystallographic analyses [59], revealed the following interesting detail. Unlike reaction of dioxo-Mo(VI)-complex (M=O) with the substrate that results in the formation of Mo(IV) species, the reaction of μ-oxo-Mo(VI)-complex (M-O-M) results in the formation of Mo(V) dimer species (Scheme **3**) which are inactive as oxygen atom transfer agents to organic compounds. They are paramagnetic species that are prone to oligomerization in reaction conditions: From these facts, it becomes clear that owing to the presence of μ-oxo (M-O-M), referred to as species 1 in the equation below, in the solution, the overall reaction of $[(O=Mo(VI)=O)X_2]L$ with a substrate (for instance, Ph_3P) may not be complete:

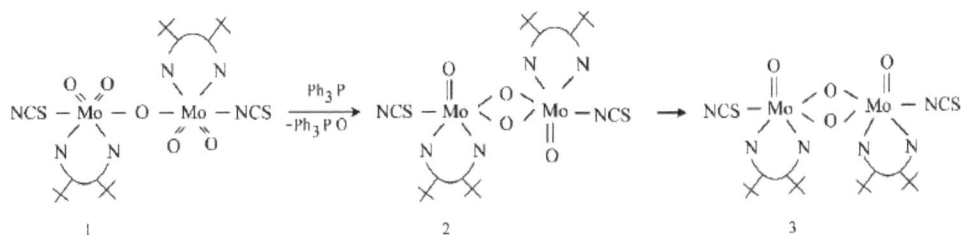

where

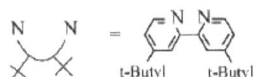

Depending on the reaction conditions, part of the initial dioxo-Mo(VI)-complex at the end even of the first reaction cycle (3) becomes unreducible by a second oxidant, for example, dioxygen. Therefore, the improvement of the oxidation process may be achieved by preventing the dimerization (oligomerisation) of the dioxo-Mo(VI)-complex in the reaction. In this regard, among the different approaches to solving this problem, the heterogenization of the dioxo-Mo(VI)-complex by grafting it onto a photosensitive solid support is one of the most effective solutions. The Mo-complex anchored on the surface of nanostructured TiO_2 (P25) is presented in Fig. **(4)**.

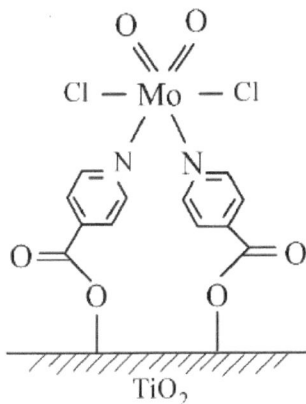

Fig. (4). Dioxomolybdenum(VI)-dichloro[4,4'-dicarboxylato-2,2'-bipyridine] complex, covalently anchored on the TiO_2 (P25).

The methods of synthesis and characterisation of this complex have been described in [60, 61]. As it has been shown later, it exhibited exceptional capacity to transfer the oxygen atom to different organic compounds under UV-irradiation [60 - 64]. From our point of view, at least two main factors determine the enhanced photoassisted reactivity of the anchored complex on TiO_2 in redox reactions [63]. The first of them is the isolation of each Mo-centre from the others through the attachment of the ligand on the surface of TiO_2 by covalent bonds. Evidently, this kind of molecular structure of the anchored complex (Fig. **4**) either prevents or at least hinders the dimerization of the oxo-Mo monomer. The second important factor is related to the redistribution of the electronic density in the anchored complex under UV-irradiation. The energy of UV irradiation absorbed by the surface layer of TiO_2, is used to create pairs of electrons and holes. Electrons diffusing to the surface can notably change the electronic density around the Mo(VI) coordination centre through the ligand, contributing to their enhanced reactivity in oxygen transfer reactions.

We investigated the photocatalytic oxidative degradation of some chlorophenyl substituted alkanes (Fig. **5**) with dioxygen in the presence of the anchored Mo-complex on TiO_2, under UV irradiation (λ= 253.7 nm) [65]. These compounds are of interest from the point of view of some problems in environmental chemistry.

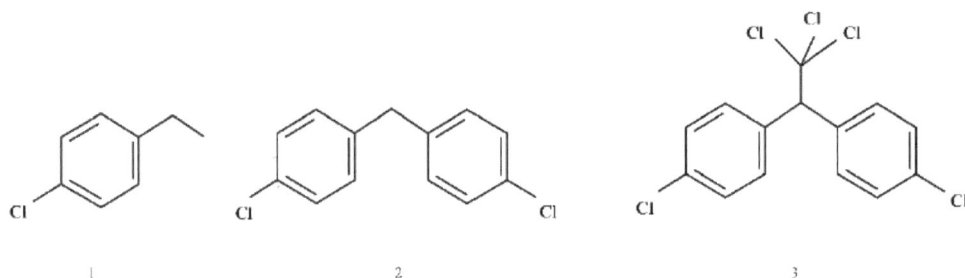

Fig. (5). (1). 1-chloro-4-ethylbenzene (CEB); (2). 4,4'-dichlorodiphenylmethane (DDM); (3). 1,1,1-trichlor--2,2-bis(p-chlorophenyl)ethane (DDT).

In the case of all three compounds, the main reaction products were corresponding alcohols formed by oxygen atom transfer reactions to the substrate. Other products of the reactions were compounds formed mainly by the decomposition of alcohols and oxidation of their products in the secondary reactions. The kinetic peculiarities of oxygen atom transfer from dioxo-Mo(VI) moieties and their reduction to oxo-Mo(IV) are shown in [63, 64].

Among these compounds, DDT (dichlorodiphenyltrichloroethane) deserves special attention because it is a persistent organic pollutant that contaminates soil and water [66]. In ordinary conditions, it may not be oxidized with air or dioxygen and even with some strong oxidants, such as chromic oxide in glacial acetic acid, and nitric acid [67]. However, it has been shown that the direct oxidation of DDT with molecular oxygen might be carried out at room temperature by applying the mentioned dioxo-Mo(VI)-complex anchored on TiO_2, under UV-irradiation in acetonitrile [66]. The overall process was carried out in two reaction periods. The first period involved the interaction of dioxo-Mo(VI)-complex with DDT under UV-irradiation, practically in the absence of dioxygen (argon atmosphere), and the second period without UV-irradiation and in the presence of dioxygen. The second period was intended to oxidize the reduced form of the complex to its initial state, in a "dark" reaction. During 4-5 experimental cycles, consisting of the two mentioned periods, about 33-35% of the initial DDT may be oxidized to chlorinated and non-chlorinated products of oxidation. The main product of the reaction was dicofol (alcohol), and most of the other products were formed as a result of further oxidation and decomposition.

Experiments on the oxidation of chlorophenyl substituted alkanes revealed some important peculiarities of oxygen atom transfer in these systems [65]. From the experimental results obtained on the oxidation of chlorophenyl alkanes, it follows that the transfer of an oxo-atom from Mo(VI) to the substrate under UV-irradiation leads to the selective oxidation of benzylic or bi-benzylic C-H bonds with the formation of the corresponding alcohols [65]. On the other hand, they indicate that the regeneration of reduced Mo(IV) to Mo(VI) occurs with dioxygen even in a "dark reaction", mainly, by the formation of oxo-peroxo-Mo(VI)-complex/ TiO$_2$ intermediates, finally producing the initial dioxo-Mo(VI)-complex TiO$_2$, according to the reaction (*III*) in the below scheme.

In the case of the oxidation of DDT [64], the following scheme of the catalytic cycle (Scheme **4**) has been suggested:

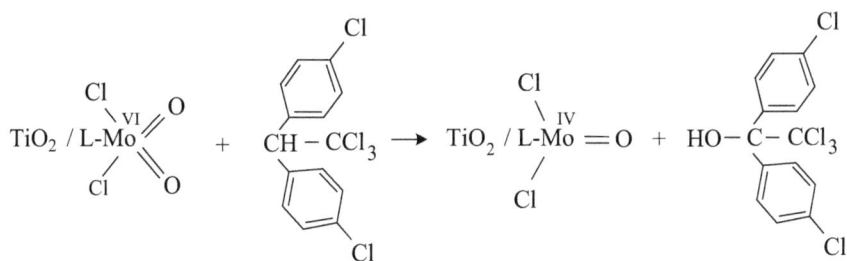

where L is 4,4'-dicarboxylato-2,2'-bipyridine

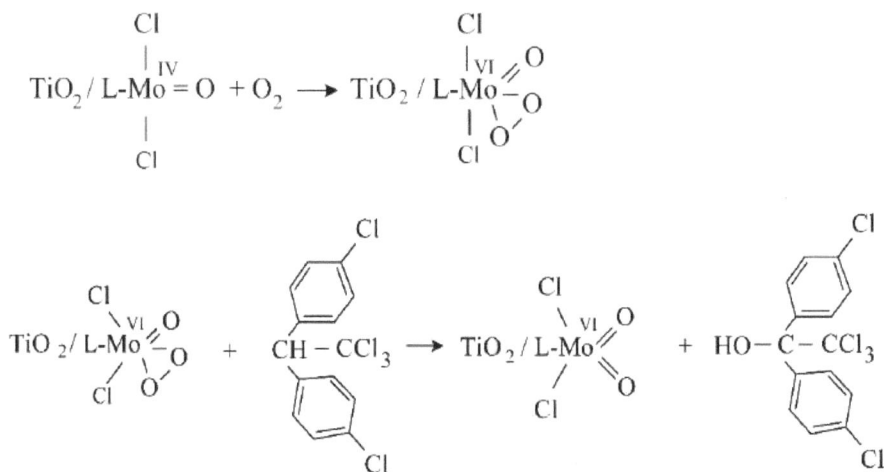

Scheme (4). Catalytic cycle of the oxidation of DDT with dioxygen catalyzed by dioxo-Mo(VI) complex anchored on TiO_2.

There are also other examples demonstrating the successful use of anchored dioxo-Mo(VI)-complexes as oxygen atom transfer agents in the oxidation reactions [68 - 70]. The application of the same complex anchored on TiO_2 nanotube, resulted in a selective epoxidation reaction (>90%) of olefins (α-pinene [69], as well as monoterpenes, such as β-pinene, camphene, (R)-(+)-limonene and (S)-(-)-limonene [70]).

Consideration and review of the above-mentioned and other experimental data show that the heterogeneous photocatalytic reaction of the transfer of oxygen atoms from the transition metal-oxo moieties to substrates can be induced in the following cases:

(1) when the process occurs by photoexcitation of transition metal-oxo compounds, or more precisely by photoexcitation of the M=O moiety, on an "inert" (non-photosensible) support material, often, resulting in the photoinduced partial electron transfer from the metal to the oxygen atom

$$M=O \xrightarrow{h\nu} M^+ - O^-$$

It is named local excitation. M^+-O^- is an active oxygen species that can easily transfer an oxygen atom to different substrates [31],

(2) when the process occurs by photoexcitation of a photosensible support material, particularly, a semiconductor, supporting an "inert" (non-photosensible) metal-oxo compound (combined mainly with chemical bonds to the surface) as a result of which an electron/hole pair appears, which in turn changes the electronic distribution in metal-oxo moieties and activates it in the oxygen transfer reaction,

(3) when photoexcitation takes place in both the metal-oxo catalyst and the support material, facilitating the transfer of oxygen atoms to substrates. However, this is a rare case.

In all these cases, the overall process can be characterized as a photoassisted catalytic reaction under heterogeneous conditions.

CONCLUSION

As a general conclusion, it is obvious that light of a certain wavelength in a heterogeneous photocatalytic system generates or induces the generation of such active oxygen species, which can promote oxygen transfer reactions in the oxidation of organic compounds.

The above-described experimental investigations of the oxidation of a great number of organic compounds (mainly with dioxygen and, often, also by other oxidants) are evidence that oxygen atom transfer reactions in heterogeneous photocatalytic systems may be either the main or one of the essential pathways in the reaction mechanism. The fundamental importance of the oxygen atom transfer reaction in the oxidation processes on the most investigated heterogeneous photocatalyst TiO_2, including in aqueous media, was shown by the application of different experimental methods, particularly the isotopic labelling technique.

According to generally accepted perceptions, the uncertainties in the experimental determination of quantum yields in heterogeneous photocatalytic systems do not allow the unambiguous characterization of the reactions of oxygen atom transfer as either *photogenerated* (photoinduced, photoinitiated) or *catalytic photolyzed* (photosensitized) processes. However, it seems that the majority of oxygen atom transfer reactions in heterogeneous photocatalytic systems are photoassisted rather than photogenerated.

The improvement of the oxygen atom transfer capacity of transition metal-oxo complexes in photocatalytic processes by heterogenization of reaction systems, grafting the catalyst onto photosensitive solid substances through chemical bonds, is one of the perspective pathways for the development of selective oxidation reactions in chemical syntheses and efficient mineralization processes in environmental chemistry.

The revelation of a detailed mechanism of oxygen atom transfer reactions opens up new opportunities for creating well-controllable photocatalytic processes corresponding to the requirements of "green" and sustainable chemistry, as well as to create appropriate models of oxygen atom transfer processes in enzymatic photocatalysis.

CONSENT FOR PUBLICATION

Not applicable.

CONFLICT OF INTEREST

The authors declare no conflict of interest, financial or otherwise.

ACKNOWLEDGEMENT

Declared none.

REFERENCES

[1]   Gisbertz, S.; Piebe, B. Heterogeneous photocatalysis in organic synthesis. *ChemPhotoChem,* **2020**, *4*(7), 456-475.
        [http://dx.doi.org/10.1002/cptc.202000014]

[2]   Shneider, J.; Bahnemann, D.; Ye, J.; Puma, G.L.; Dionysiou, D.D., Eds. *Photocatalysis. Fundamentals and Perspectives*; Royal Society of Chemistry: Cambridge, UK, **2016**.
        [http://dx.doi.org/10.1039/9781782622338]

[3]   Bakhtchadjian, R. *Bimodal Oxidation: Coupling of Heterogeneous and Homogeneous Reactions*; CRC Press,Taylor and Francis: Boca Raton, London, New York, **2019**.
        [http://dx.doi.org/10.1201/9780429295829]

[4]   Nosaka, Y.; Nosaka, A. *Indroduction to Photocatalysis. From Basic Science to Application*; Royal Society of Chemistry: Cambridge, UK, **2016**.

[5]   Albini, A.; Fagnoni, M. *Photochemically-generated Intermediates in Syntheses*; John Wiley and Sons: Hoboken, **2013**.
        [http://dx.doi.org/10.1002/9781118689202]

[6]   Ameta, R.; Chohadia, A.K.; Jain, A.; Punjabi, P.B. Fenton and Photo-Fenton Processes. In: *Advanced Oxidation Processes for Waste Water Treatment. Emerging Green Chemical Technology*; Academic Press: London, **2018**; pp. 49-87.
        [http://dx.doi.org/10.1016/B978-0-12-810499-6.00003-6]

[7]   Holm, R.H. Metal-centered oxygen atom transfer reactions. *Chem. Rev.,* **1987**, *87*(6), 1401-1449.
        [http://dx.doi.org/10.1021/cr00082a005]

[8]   Parrino, F.; Palmisano, L. Reactions in the Presence of Irradiated Semiconductors: Are They Simply Photocatalytic? *Mini Rev. Org. Chem.,* **2018**, *15*(2), 157-164.
        [http://dx.doi.org/10.2174/1570193X14666171117151718]

[9]   Linsebigler, A.L.; Lu, G.; Yates, J.T., Jr Photocatalysis on TiO$_2$ surfaces: Principles, mechanisms, and selected Results. *Chem. Rev.,* **1995**, *95*(3), 735-758.
        [http://dx.doi.org/10.1021/cr00035a013]

[10]   Lippard, S.J.; Berg, J.M. *Principles of Bioinorganic Chemistry*; University Science Book: Mill Valley,

California, **1994**, p. 318.

[11] Guo, Z.; Liu, B.; Zhang, Q.; Deng, W.; Wang, Y.; Yang, Y. Recent advances in heterogeneous selective oxidation catalysis for sustainable chemistry. *Chem. Soc. Rev.,* **2014**, *43*(10), 3480-3524.
[http://dx.doi.org/10.1039/c3cs60282f] [PMID: 24553414]

[12] Yamaguchi, K.; Mizuno, N. Scope, kinetics, and mechanistic aspects of aerobic oxidations catalyzed by ruthenium supported on alumina. *Chemistry,* **2003**, *9*(18), 4353-4361.
[http://dx.doi.org/10.1002/chem.200304916] [PMID: 14502621]

[13] Krumova, K.; Cosa, G. Overview of Reactive Oxygen Species. In: *Singlet Oxygen: Applications in Biosciences and Nanosciences (Comprehensive Series in Photochemical & Photobiological Sciences. V.1)*; Royal Society of Chemistry: Cambridge, UK, **2016**; pp. 1-21.
[http://dx.doi.org/10.1039/9781782622208-00001]

[14] Sankaralingam, M.; Lee, Y-M.; Nam, W.; Fukuzumi, S. Amphoteric reactivity of metal–oxygen complexes in oxidation reactions. *Coord. Chem. Rev.,* **2018**, *365*, 41-59.
[http://dx.doi.org/10.1016/j.ccr.2018.03.003]

[15] Neubert, S.; Mitoraj, D.; Shevlin, S.A.; Pulisova, P.; Heimann, M.; Du, Y.; Goh, G.K.L.; Pacia, M.; Kruczała, K.; Turner, S.; Macyk, W.; Guo, Z.X.; Hocking, R.K.; Beranek, R. Guo; Z. X.; Hocking, R.; Beranek, R. Highly efficient rutile TiO_2 photocatalysts with single Cu(II) and Fe(III) surface catalytic sites. *J. Mater. Chem. A Mater. Energy Sustain.,* **2016**, *4*(8), 3127-3138.
[http://dx.doi.org/10.1039/C5TA07036H]

[16] Nosaka, A.Y.; Nosaka, Y. Understanding hydroxyl radical (•OH) generation processes in photocatalysis. *ACS Energy Lett.,* **2016**, *1*(2), 356-359.
[http://dx.doi.org/10.1021/acsenergylett.6b00174]

[17] Murakami, Y.; Kenji, E.; Nosaka, A. Y.; Nosaka, Y. \ Direct detection of OH radicals defused to the gas phase from the UV irradiation photocatalytic TiO_2 surface by means laser-indused fluorescence spectroscopy. *Journal Phys. Chem.,* **2006**, *110, 34, 16*, 808-16 811.

[18] Zhang, J.; Nosaka, Y. Mechanism of the OH radical generation in photocatalysis with TiO_2 of different crystalline types. *J. Phys. Chem. C,* **2014**, *118*(20), 10824-10832.
[http://dx.doi.org/10.1021/jp501214m]

[19] Coronado, J.M.; Maira, A.J.; Conesa, J.C.; Yeung, K.L.; Augugliaro, V.; Soria, J. EPR study of the surface characteristics of nanostructured TiO_2 under UV Irradiation. *Langmuir,* **2001**, *17*(17), 5368-5374.
[http://dx.doi.org/10.1021/la010153f]

[20] Yi, J.; Bahrini, C.; Schoemaecker, C.; Fittschen, C.; Choi, W. Photocatalytic decomposition of H_2O_2 on different TiO_2 surfaces along with the concurrent generation of HO_2 radicals monitored using cavity ring down spectroscopy. *J. Phys. Chem. C,* **2012**, *116*(18), 10090-10097.
[http://dx.doi.org/10.1021/jp301405e]

[21] Kakuma, Y.; Nosaka, A.Y.; Nosaka, Y. Difference in TiO_2 photocatalytic mechanism between rutile and anatase studied by the detection of active oxygen and surface species in water. *Phys. Chem. Chem. Phys.,* **2015**, *17*(28), 18691-18698.
[http://dx.doi.org/10.1039/C5CP02004B] [PMID: 26120611]

[22] Lin, Y.; Deng, C.; Wu, L.; Zhang, Y.; Chen, C.; Ma, W.; Zhao, J.; Zhao, J. Quantitative isotope measurements in heterogeneous photocatalysis and electrocatalysis. *Energy Environ. Sci.,* **2020**, *13*(9), 2602-2617.
[http://dx.doi.org/10.1039/D0EE01790F]

[23] Zhang, M.; Wang, Q.; Chen, C.; Zang, L.; Ma, W.; Zhao, J. Oxygen atom transfer in the photocatalytic oxidation of alcohols by TiO_2: oxygen isotope studies. *Angew. Chem. Int. Ed. Engl.,* **2009**, *48*(33), 6081-6084.
[http://dx.doi.org/10.1002/anie.200900322] [PMID: 19343745]

[24]   Pang, X.; Chen, C.; Ji, H.; Che, Y.; Ma, W.; Zhao, J. Unravelling the photocatlytic mechanisms on TiO₂ surfaces using oxygen -18 isotopic label technique In: *Molecules*; , **2016**; pp. 52-72.

[25]   Pang, X.; Chang, W.; Chen, C.; Ji, H.; Ma, W.; Zhao, J. Determining the TiO2-photocatalytic aryl-ring-opening mechanism in aqueous solution using oxygen-18 labeled O2 and H2O. *J. Am. Chem. Soc.,* **2014**, *136*(24), 8714-8721.
[http://dx.doi.org/10.1021/ja5031936] [PMID: 24850419]

[26]   Lang, X.; Ma, W.; Chen, C.; Ji, H.; Zhao, J. Selective aerobic oxidation mediated by TiO₂ photocatalysis. *Acc. Chem. Res.,* **2014**, *47*(2), 355-363.
[http://dx.doi.org/10.1021/ar4001108] [PMID: 24164388]

[27]   Lang, X.; Chen, C.; Ma, W.; Ji, H.; Zhao, J. Oxygen-atom transfer in titanium dioxide photoredox catalysis for organic synthesis (chapter). *Photochemistry,* **2017**, *44*, 364-384.
[http://dx.doi.org/10.1039/9781782626954-00364]

[28]   Agasti, N. TiO2 nanomaterials for photocatalytic application. Chapter 12. In: *Smart Ceramics, Preparation, properties and application*; Mishra, A.K. Pan Stanford Publishing Pte Ltd: Singapore, **2018**.

[29]   Chen, C.; Wang, Z.; Che, Y.; Ma, W.; Ji, H.; Zhao, J. Photocatalytic degradation of organic contaminants on mineral surfaces. In: *Biophysico-Chemical Processes of Anthropogenic Organic Compounds in Environmental Systems. IUPAC Series*; Xing, B.; Senesi, N.; Huang, P.M. Wiley: Hoboken, **2011**.
[http://dx.doi.org/10.1002/9780470944479.ch4]

[30]   Salomon, R.G. Homogeneous metal-catalysis in organic photochemistry. *Tetrahedron,* **1983**, *39*(4), 485-575.
[http://dx.doi.org/10.1016/S0040-4020(01)91830-7]

[31]   Serpone, N.; Emeline, A.V. Suggested terms and definitions in photocatalysis and radiocatalysis. *Int. J. Photoenergy,* **2002**, *4*(3), 91-131.
[http://dx.doi.org/10.1155/S1110662X02000144]

[32]   Serpone, N.; Emeline, A.V. Fundamentals in metal oxide heterogeneous catalysis. In: *Nanostructured and Electrochemical Systems For Solar Photon Conversion*; Archer, M.D.; Nozik, A.J.; Xin, A., Eds.; Imperial College Press: London, **2008**; pp. 275-381.
[http://dx.doi.org/10.1142/9781848161542_0005]

[33]   Braslavsky, S.E.; Braun, A.M.; Cassano, A.E.; Emeline, A.V.; Litter, M.I.; Palmisano, L.; Parmon, V.N.; Serpone, N. Glossary of terms used in photocatalysis and radiation catalysis (IUPAC Recommendations). *Pure Appl. Chem.,* **2011**, *83*(4), 931-1014.
[http://dx.doi.org/10.1351/PAC-REC-09-09-36]

[34]   Ohtani, B. Photocatalysis A to Z —What we know and what we don't know in a scientific sense. *J. Photochem. Photobiol. Photochem. Rev.,* **2010**, *11*(4), 157-178.
[http://dx.doi.org/10.1016/j.jphotochemrev.2011.02.001]

[35]   Cismesia, M.A.; Yoon, T.P. Characterizing chain processes in visible light photoredox catalysis. *Chem. Sci. (Camb.)* **2015**, *6*(10), 5426-5434.
[http://dx.doi.org/10.1039/C5SC02185E] [PMID: 26668708]

[36]   Kisch, H. On the problem of comparing rates or apparent quantum yields in heterogeneous photocatalysis. *Angew. Chem. Int. Ed. Engl.,* **2010**, *49*(50), 9588-9589.
[http://dx.doi.org/10.1002/anie.201002653] [PMID: 21077070]

[37]   Serpone, N. Relative photonic efficiencies and quantum yields in heterogeneous photocatalysis. *Journal of Photochemistry and Photobiology A: Chemistry.,* **1997**, *104*, 1.3 1-12.

[38]   Kisch, H.; Bahnemann, D. Best practice in photocatalysis: comparing rates or apparent quantum yields? *J. Phys. Chem. Lett.,* **2015**, *6*(10), 1907-1910.
[http://dx.doi.org/10.1021/acs.jpclett.5b00521] [PMID: 26263267]

[39]   Hoque, M.A.; Guzman, M.I. Photocatalytic activity: Experimental features to report in heterogeneous Photocatalysis. *Materials (Basel)*, **2018**, *11*(10), 1990.
[http://dx.doi.org/10.3390/ma11101990] [PMID: 30326644]

[40]   Machuca, F.; Colina–Márquez, J.; Mueses, M. Determination of quantum yield in a heterogeneous photocatalytic system using a fitting-parameters model. *J. Adv. Oxid. Technol.*, **2008**, *11*(1), 42-49.
[http://dx.doi.org/10.1515/jaots-2008-0105]

[41]   Wang, C.; Bahnemann, D.W.; Dohrman, J.K. Determination of photonic efficiency and quantum yield of formaldehyde formation in the presence of various TiO2 photocatalysts. *Water Sci. Technol.*, **2001**, *44*(5), 279-286.
[http://dx.doi.org/10.2166/wst.2001.0306] [PMID: 11695471]

[42]   Reiß, B. Hu, Q.; Riedle, E.; Wagenknecht, H-A. Dependence of Chemical Quantum Yields of Visible Light Photoredox Catalysis on the Irradiation Power. *Chem Photo Chem*, **2021**, *5*(11), 1009-1019.

[43]   Sánchez-Velandia, J.E.; Páez-Mozo, E.A.; Martínez, F.O. A kinetic study of the photoinduced oxo-transfer using a Mo complex anchored to TiO$_2$. *Rev. Fac. Ing. Univ. Antioquia*, **2021**, *98*, 83-93.

[44]   Oppenlander, T. Photochemical processes of water treatment. In: *Photochemical Purification of Water and Air: Advanced Oxidation Processes (AOPs) - Principles, Reaction Mechanisms, Reactor Concepts*; Wiley-VCH: Weinheim, **2003**; pp. 101-144.

[45]   Eves, B.J.; Lopinski, G.P. Formation of organic monolayers on silicon *via* gas-phase photochemical reactions. *Langmuir*, **2006**, *22*(7), 3180-3185.
[http://dx.doi.org/10.1021/la052960a] [PMID: 16548575]

[46]   Chatgilialoglu, C.; Timokhin, I.V. Silile radicals in chemical syntheses.*Advances in Organometallic Chemistry*; Hill, A.F.; Fink, M.J., Eds.; Elsevier: Amsterdam, **2008**, Vol. 57, pp. 117-182.

[47]   Richards-Henderson, N.K.; Goldstein, A.H.; Wilson, K.R. Large enhancement in the heterogeneous oxidation rate of organic aerosols by hydroxyl radicals in the presence of nitric oxide. *J. Phys. Chem. Lett.*, **2015**, *6*(22), 4451-4455.
[http://dx.doi.org/10.1021/acs.jpclett.5b02121] [PMID: 26505970]

[48]   Valange, S.; Védrine, J.C. General and prospective views on oxidation reactions in heterogeneous catalysis. *Catalysts*, **2018**, *8*(10), 483. [Review].
[http://dx.doi.org/10.3390/catal8100483]

[49]   a) Barona-Castaño, J.C.; Carmona-Vargas, C.C.; Brocksom, T.J.; de Oliveira, K.T. Porphyrins as catalysts in scalable organic reactions. *Molecules*, **2016**, *21*(3), 310.
[http://dx.doi.org/10.3390/molecules21030310] [PMID: 27005601] b) Verho, O. *Metal Catalyzed Redox Reactions*, Doctoral thesis, Stockholm University. **2013**,

[50]   Mak, C.H.; Han, X.; Du, M.; Kai, J-J.; Tsang, K.F.; Jia, G.; Cheng, K-C.; Shen, H.H.; Hsu, H-Y. Heterogenization of homogeneous photocatalysts utilizing synthetic and natural support materials. *J. Mater. Chem. A Mater. Energy Sustain.*, **2021**, *9*(8), 4454-4504.
[http://dx.doi.org/10.1039/D0TA08334H]

[51]   Alvaro, M.; Carbonell, E.; Espla, M.; Garcia, H. Ironphthalocyanine supported on silica or encapsulatedinside zeolite Y as solid photocatalysts for thedegradation of phenols and sulfur heterocycles. *Appl. Catal. B*, **2005**, *57*(1), 37-42.
[http://dx.doi.org/10.1016/j.apcatb.2004.10.003]

[52]   Su, H.; Li, Z.; Huo, Q.; Guan, J.; Kan, Q. Immobilization of transition metal (Fe^{2+}, Co^{2+}, V^{2+}, Cu^{2+}) Schiff base onto graphene oxide as efficient and recyclable catalysts for epoxidation of styrene. *RSC Advances*, **2014**, *4*(20), 9990-9996.
[http://dx.doi.org/10.1039/c3ra47732k]

[53]   Ahadi, E.; Hosseini-Monfared, H.; Spieß, A.; Janiak, C. Photocatalytic asymmetric epoxidation of *trans*-stilbene with manganese–porphyrin/graphene-oxide nanocomposite and molecular oxygen: axial

ligand effect. *Catal. Sci. Technol.,* **2020**, *10*(10), 3290-3302.
[http://dx.doi.org/10.1039/D0CY00441C]

[54]   Ciamician, G.; Silber, P. Chemische Lichtwirkungen. *Ber. Dtsch. Chem. Ges.,* **1901**, *34*(2), 2040-2046.
[http://dx.doi.org/10.1002/cber.190103402118]

[55]   Hille, R.; Rétey, J.; Bartlewski-Hof, U.; Reichenbecher, W.; Schink, B. Mechanistic aspects of molybdenum-containing enzymes. *FEMS Microbiol. Rev.,* **1998**, *22*(5), 489-501.
[http://dx.doi.org/10.1111/j.1574-6976.1998.tb00383.x] [PMID: 10189201]

[56]   Arzoumanian, H. Molybdenum-oxo chemistry in various aspects of oxygen atom transfer. *Coord. Chem. Rev.,* **1998**, *178–180*(1), 191-202.
[http://dx.doi.org/10.1016/S0010-8545(98)00056-3]

[57]   Arzoumanian, H.; Bakhtchadjian, R. Oxo-atom transfer reactions of transition metal complexes in catalytic oxidation with O2 on the light of some recent results in molybdenum-oxo chemistry (a review). *Chemical Journal of Armenia,* **2012**, *65*(2), 168-188.

[58]   Arzoumanian, H.; Bakhtchadjian, R.; Agrifogio, G.; Krentzien, H.; Daran, J-C. Synthesis and characterizations of a dioxo-μ-oxo molybdenum dimer: An unusual case of a μ-oxo conformational equilibrium. *Eur. J. Inorg. Chem.,* **1999**, *1999*(12), 2255-2259.
[http://dx.doi.org/10.1002/(SICI)1099-0682(199912)1999:12<2255::AID-EJIC2255>3.0.CO;2-N]

[59]   Arzoumanian, H.; Bakhtchadjian, R.; Atencio, R.; Briceno, A.; Agrifolio, G. Characterization of a reduced molybdenum-oxo compound derived from an oxo-transfer process under stoichiometric conditions. *J. Mol. Catal. Chem.,* **2006**, *260*(1-2), 1-2, 197-206.
[http://dx.doi.org/10.1016/j.molcata.2006.07.025]

[60]   Arzoumanian, H.; Castellanos, N.J.; Martınez, F.O.; Paez-Mozo, E.A.; Ziarelli, F. Silicon-assisted direct covalent grafting on metal oxide surfaces: Synthesis and Characterization of carboxylate N,N′-ligands on TiO$_2$. *Eur. J. Inorg. Chem.,* **2010**, *11*(11), 1633-1641.
[http://dx.doi.org/10.1002/ejic.200901092]

[61]   Bakhtchadjyan, R.; Tsarukyan, S.V.; Manucharova, L.A.; Tavadyan, L.A.; Barrault, J.; Matinez, F.O. Photochemical oxidative decomposition of 1-chloro-4-ethylbenzene in the presence of dioxo-molybdenum(VI) complex anchored on the TiO2. *Chemical Journal of Armenia,* **2011**, *64*(1), 9-15.

[62]   Castellanos, N.J.; Martınez, F.; Lynen, F.; Biswas, S.; Van Der Voort, P.; Arzoumanian, H. Dioxygen activation in photooxidation of diphenylmethane by a dioxo- molybdenum(VI) complex anchored covalently onto mesoporous titania. *Trans. Met. Chem,* **2013**, *38*(2), 119-127.
[http://dx.doi.org/10.1007/s11243-012-9668-2]

[63]   Bakhtchadjyan, R.A.; Tsarukyan, S.V.; Manucharova, L.A.; Tavadyan, L.A.; Barrault, J.; Matinez, F.O. Photochemical decomposition of 1- chloro-4-ethylbenzene by participation of O$_2$ and a dioxomolybdenum(VI) complex anchored on the TiO$_2$ surface. *Kinet. Catal.,* **2013**, *54*(1), 34-39.
[http://dx.doi.org/10.1134/S0023158413010011]

[64]   Bakhtchadjyan, R.; Manucharova, L.A.; Tavadyan, L.A. Selective oxidation of DDT by dioxygen on the dioxo- Mo(VI) complex anchored on a TiO$_2$ under UV-irradiation. *Catal. Commun.,* **2015**, *69*, 193-195.
[http://dx.doi.org/10.1016/j.catcom.2015.06.016]

[65]   Bakhtchadjyan, R.; Manucharova, L.A.; Tavadyan, L.A. Organometallic Mo(VI)-complex Grafted on TiO$_2$ as Photocatalyst in Oxidation of Chlorophenyl Substituted Alkanes with Dioxygen. In: *Advances in Chemistry Research*; Vol. 57; Taylor, J.C., Ed.; Nova Science Publishers: N.Y., **2020**; pp. 163-190.

[66]   Bakhtchadjyan, R.; Manucharova, L.A.; Tavadyan, L.A. Photocatalytic Selective Oxidation of DDT(dichlorodiphenyltrichloroethane) to Dicofol. In: *DDT: Properties, Uses and Toxicity. Series: Environmental Remediation Technologies, Regulations and Safety*; Sanders, K., Ed.; Nova Science Publishers: N.Y., USA, **2016**; pp. 85-106.

[67]   Lawless, E.W.; Ferguson, T.L.; Meiners, A.F. Guidelines for the Disposal of Small Quantities of Unused Pesticides, Part A. Pesticides, Pesticides Chemistry, and Pesticide Disposal (Section 7. Review of the chemistry of pesticide disposal (pp 53-134) and DDT (pp.104-105). In: *Environmental Protection Agency*; National Environmental Research Centre, Office of Research and Development: Cincinnati, OH, **1975**.

[68]   Biswas, S.; Van Der Voort, P.; Castellanos, N.J.; Martínez, F. Dioxygen activation in photooxidation of diphenylmethane by a dioxomolybdenum(VI) complex anchored covalently onto mesoporous titania. *Trans. Met. Chem.,* **2013**, *38*(2), 119-127.
[http://dx.doi.org/10.1007/s11243-012-9668-2]

[69]   Henry, M.; Álvaro, A.A.; Páez-Mozo, E.A.; Martinez, F.O. Highly efficient epoxidation of α-pinene with O_2 photocatalyzed by dioxoMo(VI) complex anchored on TiO_2 nanotube. *Microporous Mesoporous Mater.,* **2018**, *265*, 202-210.
[http://dx.doi.org/10.1016/j.micromeso.2018.02.005]

[70]   Henry, M.; Álvaro, A. A.; Páez-Mozo, E. A.; Martinez, F.O.; Valange, S. Photo-assisted O-atom transfer to monoterpenes with molecular oxygen and a dioxoMo(VI) complex immobilized on TiO2 nanotubes. *Catalysis Today,* **2021**, *475*(1), 445-457.

SUBJECT INDEX

square pyramidal _16, 65, 67 74, 82

H

Heme-iron 40, 41, 43, 46, 52
Heterogenization 103, 106,110
High-valent 4, 5, 18, 19, 21, 63, 66,
           74, 76, 79-83
Homolysis 62, 68, 72, 74, 82, 83
Hydrogen bond network 41, 52 -55
Hydroperoxide 23, 53, 54, 63, 64
Hydroxylation
   at the benzylic position 67
   reaction 42, 52, 63
Hydroxylation of
   alkane 63,70, 74
   aromatic ring 68, 74, 98
   cyclohexane 70, 82
   ligand 74, 82
   methane 43, 45
   polyethylene 78, 79

I

Intermediate
   active oxygen_ 65, 72
   cage 20
   cation radical_ 41
   iron-oxo_ 17
   nickel-dioxygen_ 65, 67, 80
    oxo-peroxo 19
   Q 43-45
   radical_ 47, 51, 55
   reaction_ 92
   superoxide radical_ 48, 54, 55
   three-center_ 17

K

KIE-Kinetic isotope effect 19, 75, 76

L

Ligand
   alkoxide_ 68
   amido_ 50
   macrocyclic_ 65
   oxo_ 3, 15, 16, 18
   porphyrin_ 64
   pyridylalkylamine__71
   pyridine_ 13
   tetradentate _71, 72
   tripodal 70-72, 80

M

Mechanism
   oxygen atom transfer_ 18-20(see
          also Oxygen)
   oxygen rebound_ 18, 19, 30, 42
   radical_ 30, 80, 82
   reaction_ 5-7, 9, 10, 23,
         27, 33, 110
   Mars-van-Krevelen_11
   water oxidation_ 12, 22-24
Mechanism of
   dioxygenation reaction 46
   dioxygen activation by
         dioxygenase 46
         Lewis acid promoted 55
         sMMO and pMMO 42, 43
         P450 40
   chain-radical reaction 9, 31, 102
   hydrogen atom transfer 77
   Wacker process 9
Method 101
   EPR_ 96
   of fluorescence spectroscopy 96
   of primary kinetic isotope effect 19
   NMR_ 105
   UV-spectroscopic_105
Mn(II), Mn(III) 43, 64, 69
Monooxygenase(s) 39-46, 51, 57
Mo(IV), Mo(V) 104, 105, 108
Mo(VI) 101, 106-109,

www.ingramcontent.com/pod-product-compliance
Lightning Source LLC
Chambersburg PA
CBHW041445210326
41599CB00004B/143